DATE DUE

			PRINTED IN U.S.A.

About Island Press

Since 1984, the nonprofit Island Press has been stimulating, shaping, and communicating the ideas that are essential for solving environmental problems worldwide. With more than 800 titles in print and some 40 new releases each year, we are the nation's leading publisher on environmental issues. We identify innovative thinkers and emerging trends in the environmental field. We work with world-renowned experts and authors to develop cross-disciplinary solutions to environmental challenges.

Island Press designs and implements coordinated book publication campaigns in order to communicate our critical messages in print, in person, and online using the latest technologies, programs, and the media. Our goal: to reach targeted audiences—scientists, policymakers, environmental advocates, the media, and concerned citizens—who can and will take action to protect the plants and animals that enrich our world, the ecosystems we need to survive, the water we drink, and the air we breathe.

Island Press gratefully acknowledges the support of its work by the Agua Fund, Inc., The Margaret A. Cargill Foundation, Betsy and Jesse Fink Foundation, The William and Flora Hewlett Foundation, The Kresge Foundation, The Forrest and Frances Lattner Foundation, The Andrew W. Mellon Foundation, The Curtis and Edith Munson Foundation, The Overbrook Foundation, The David and Lucile Packard Foundation, The Summit Foundation, Trust for Architectural Easements, The Winslow Foundation, and other generous donors.

The opinions expressed in this book are those of the author(s) and do not necessarily reflect the views of our donors.

VITAL SIGNS
2012

VITAL SIGNS

2012

The Trends That Are Shaping Our Future

WORLDWATCH INSTITUTE

Michael Renner, *Project Director*

Ayodeji Adebola	Matt Lucky
E. L. Beck	Haibing Ma
Will Bierbower	Elizabeth Leahy Madsen
Jesse Chang	Jorge Moncayo
Robert Engelman	Natalie Narotzky
Farhad Farahmand	Bo Normander
Gary Gardner	Alexander Ochs
Matthias Kimmel	Michael Renner
Saya Kitasei	Sam Shrank
Annette Knödler	Matt Styslinger
Mark Konold	Richard H. Weil

Linda Starke, *Editor*
Lyle Rosbotham, *Designer*

 **ISLAND**PRESS

Washington | Covelo | London

Contents

9 **Acknowledgments**

11 **Preface**

15 **Energy Trends**

Oil Market Resumes Growth after Stumble in 2009 16

Global Natural Gas Consumption Regains Momentum 19

Nuclear Generation Capacity Falls 23

Global Wind Power Growth Takes a Breather in 2010 26

Another Record Year for Solar Power, but Clouds on the Horizon 29

Biofuels Regain Momentum 32

Global Hydropower Installed Capacity and Use Increase 35

Energy Poverty Remains a Global Challenge for the Future 38

43 **Transportation Trends**

Auto Industry Stages Comeback from Near-Death Experience 44

High-Speed Rail Networks Expand 48

53 **Environment and Climate Trends**

Carbon Markets Struggle to Maintain Momentum 54

Carbon Capture and Storage Attracts Government Attention 58

61 **Food and Agriculture Trends**

World Grain Production Down in 2010, but Recovering 62

Organic Agriculture Sustained through Economic Crisis 65

Sugar Production Dips 69

Fish Production from Aquaculture Rises While Marine Fish Stocks Continue to Decline 72

Meat Production and Consumption Continue to Grow 76

79 **Global Economy and Resources Trends**

World's Forests Continue to Fall as Demand for Food and Land Goes Up 80

Tropical Forests Push Payments for Ecosystem Services onto the Global Stage 83

Value of Fossil Fuel Subsidies Declines, National Bans Emerging 86

Energy Intensity Is Rising Slightly 90

93 **Population and Society Trends**

World Labor Force Growing at Divergent Rates 94

Women Slowly Close Gender Gap with Men 98

Numbers of Overweight on the Rise 101

105 **Notes**

127 **The Vital Signs Series**

Acknowledgments

Like the earlier volumes, this edition of *Vital Signs* was made possible by the efforts of a large and diverse group of people. In addition to authors and reviewers, critical assistance was provided by colleagues who offered their expertise for editing and layout, as well as preparing each article for posting on the companion online site, at vitalsigns.worldwatch.org, and doing media outreach.

Many individual and institutional funders, as well as our exceedingly generous Board, provide the support without which our work would not be possible. Caroline Gabel and the Shared Earth Foundation have a long history as supporters of Worldwatch's work, including dedicated support for *Vital Signs Online*. Our appreciation for support during 2011 also goes to the Ray C. Anderson Family Foundation, Inc., The Bill & Melinda Gates Foundation, Barilla Center for Food & Nutrition, Climate Development and Knowledge Network (CDKN), Compton Foundation, Inc., Del Mar Global Trust, Ministry of Foreign Affairs of the Government of Finland, International Climate Initiative and the Transatlantic Climate Bridge of the German Federal Ministry for the Environment, Nature Protection and Nuclear Safety (BMU), Renewable Energy Policy Network for the 21st Century (REN21), The David B. Gold Foundation, Richard and Rhoda Goldman Fund and the Goldman Environmental Prize, Greenaccord International Secretariat, Energy and Environment Partnership with Central America (EEP),

Hitz Foundation, Institute of International Education, Inc., Steven C. Leuthold Family Foundation, MAP Royalty Inc. Sustainable Energy Education Fellowship Program, The Shared Earth Foundation, Shenandoah Foundation, Small Planet Fund of RSF Social Finance, V. Kann Rasmussen Foundation, United Nations Population Fund, Wallace Global Fund, Weeden Foundation, and the Winslow Foundation.

This edition was written by a team of 21 researchers. In addition to outside contributors E. L. Beck, Elizabeth Leahy Madsen, and Richard H. Weil, a group of veteran Worldwatch researchers and capable interns probed the globe's pulse. They include Will Bierbower, Jesse Chang, Robert Engelman, Farhad Farahmand, Gary Gardner, Matthias Kimmel, Saya Kitasei, Annette Knödler, Mark Konold, Matt Lucky, Haibing Ma, Jorge Mancayo, Natalie Narotzky, Alexander Ochs, Sam Shrank, and Matt Styslinger. From Copenhagen, Bo Normander, director of Worldwatch Institute Europe, also contributed.

Vital Signs authors receive help from reviewers and other experts who offer insights on the trends we follow. We give particular thanks this year to Colin Couchman, Walter Falcon, Brian Halweil, Carl Haub, Daniel Kandy, Lindsay Nauen, Michael Ratner, and Janet Sawin.

Vital Signs would be a cacophony of individual voices and styles were it not for the guiding hand of our trusted editor, Linda Starke. Linda has become the most steady presence in the 35-year history of Worldwatch, editing not only all the *Vital Signs* books but also every single edition of its sister publication, *State of the World*, as well as a bevy of other publications. Similarly, Lyle Rosbotham, our veteran graphic designer, plays an indispensable role in selecting the book's cover art and layout.

The year 2011 was one of transition at Worldwatch Institute, and the fact that our work continued without missing a beat is testament to the collegial spirit and hard work of various individuals. At the helm of the organization, Robert Engelman took over from Christopher Flavin. I am particularly grateful for the foresight of my predecessor as *Vital Signs* Project Director, Gary Gardner, who left in April. His careful planning ensured that the transition in responsibility came off without a hitch. Gary and I collaborated on many projects since he joined Worldwatch in July 1994, and his departure was a big loss to me and the entire organization.

We also experienced some transitions with our Web and communications staff. Bernard Pollack took over when first Colleen Kredell and later Heather Risley departed, injecting his unique brand of energy, enthusiasm, and vision.

A cadre of staff works behind the scenes to ensure that our work is funded and distributed and that the office is well managed. We thank Barbara Fallin, Mary Redfern, Patricia Shyne, and Trudy Loo (who decided to return to her native Vancouver in the spring of 2011). Special thanks go to Patricia, who is retiring after seven years of managing all of our publications. In yet another transition, Alex Kostura left the Institute in the summer of 2011 and was replaced in his roles as development associate and executive assistant by the equally capable Grant Potter.

This is the first volume in the *Vital Signs* series that is being published by Island Press, but we look forward to many years of a productive partnership with our colleagues there, especially David Miller, Brian Weese, Maureen Gately, Jaime Jennings, and Sharis Simonian.

Enjoy *Vital Signs 2012*, and stay up to date on the latest sustainability trends throughout the year at vitalsigns.worldwatch.org.

Michael Renner
Project Director
Worldwatch Institute
1776 Massachusetts Avenue, N.W.
Washington, DC 20036
vitalsigns.worldwatch.org

Preface

Judging by the international sustainability-related events that will be taking place during 2012, this will be a momentous year. The website for "Sustainable Development Conferences Worldwide" lists 262 events that are set to occur this year, though even that impressive number seems certain to be an incomplete count. Some efforts stand out. The United Nations declared 2012 to be the International Year of Sustainable Energy Access for All—which recognizes that while the wealthy and middle class of the world enjoy the benefits and comforts afforded by the combustion of fossil fuels, as many as 1.4 billion poor people have no access to electricity at all. More than 2.5 billion people rely on cooking fuels—wood and low-quality petroleum products—that produce unhealthy indoor fumes.

In June, the premier sustainability event will take place in Rio de Janeiro, where government and civil society leaders gather for the United Nations Conference on Sustainable Development. Rio 2012 marks the twentieth anniversary of the 1992 Earth Summit, a conference that parented the Framework Convention on Climate Change. Since then, 17 annual meetings on that treaty, known as "conferences of the parties" (COPs), have taken place—from Berlin in 1995 to Durban in 2011. In 2012, COP 18 will be hosted by Qatar.

Environmental terminology is in vogue. A Google search returns 170 million results for "climate conference," 116 million for "sustainability," 52.9 million for "sustainable develop-ability," ment," and 29.8 million for "green economy." This explosion of Web content has left developments in the real world behind. As this edition of *Vital Signs* shows (and this is the nineteenth year in which this book appears, and the third year since we inaugurated a companion online portal), most key trends continue to point to an unsustainable future, though there are glimmers of hope.

Global energy intensity, defined as worldwide total energy consumption divided by gross world product, decreased by just over 20 percent from 1981 to 2010—about 0.79 percent each year. Though 2010 bucked this trend—energy intensity rose 1.35 percent, largely in response to economic stimulus efforts that were more concerned with "priming the pump" than with sustainability—the overall trend is clearly positive.

At 37 percent of primary energy use worldwide, oil remains the largest single source of energy, though its share has declined for 11 consecutive years. (Natural gas, by comparison, accounts for about 24 percent.) In 2010, global oil consumption reached an all-time high of 87.4 million barrels. Fossil fuel consumption has long been supported with generous subsidies. Consumption subsidies fell 44 percent in 2009, to $312 billion—a reduction due more to changes in international energy prices than to actual curtailment of subsidies. Production subsidies are estimated at an additional $100 billion per year. Fossil fuel subsidies dwarf support for renewable energy, which totaled $47 billion in 2007.

Hydropower accounted for a relatively small 3.4 percent of global energy consumption in 2010, but for 16.1 percent of electricity consumption. Total hydropower use has risen about 3.5-fold since 1965, though often driven by destructive large-scale dams. Newer types of renewable energy are growing fast. Global wind power capacity increased 24 percent to 197,000 megawatts in 2010. Total installed capacity is now nine times larger than a decade ago, with China, the United States, Germany, and Spain the global leaders. Meanwhile, some 16,700 megawatts of solar photovoltaic (PV) generating capacity was installed in 2010, a figure that surpasses the total existing PV capacity as recently as 2008. The total installed capacity of close to 40,000 megawatts is capable of producing enough electricity to power about 13 million households. Germany alone accounts for more than 40 percent of global installed capacity.

The transportation sector is a major consumer of fossil fuels. The world auto industry staged a dramatic comeback from 2008 and 2009, when production and sales took a precipitous plunge. In 2010, production of passenger cars and so-called light trucks shot up from 60 million to a new record of 74.7 million. Sales increased to 75.4 million.

Among the alternatives to heavy car reliance, high-speed rail (HSR) is gathering steam. The length of HSR tracks worldwide grew from 10,700 kilometers in 2009 to almost 17,000 kilometers in 2011. Tracks currently under construction and planned will bring the total to close to 43,000 kilometers. HSR travel volume expanded from about 30 billion passenger-kilometers in 1970 to at least 200 billion today (a figure that does not include the fast-growing Chinese market). The high-speed segment accounted for 7 percent of all rail passenger travel in 2010.

Food and fuel uses are increasingly intertwined, with trade-offs looming large. Sugar crops (mostly sugarcane) were once largely used for food but are increasingly turned into ethanol for cars, as happens in Brazil. A rising portion of the corn produced in the United States is also put to similar uses—where ethanol was pro-

jected to overtake the animal feeding industry as the largest corn consumer in 2011. Global biofuel production (ethanol and biodiesel) increased by 17 percent in 2010 to reach an all-time high of 105 billion liters. This is up from less than 20 billion liters a decade earlier. Biofuels provided 2.7 percent of all global fuels for road transportation.

Whether destined to be food, fuel, or feed, agricultural production has been pushed higher and higher—thanks to high-yield crop varieties, to heavy reliance on irrigation, synthetic fertilizers, and pesticides, and to the attendant environmental degradation and resource depletion. Between 1961 and 2010, the world's grain harvest more than tripled, from 643 million to 2.2 billion tons, with the peak year to date in 2008.

Organic farming practices are spreading—practiced on 37.2 million hectares worldwide in 2009, a 150 percent increase since 2000. Yet the organic area is still exceedingly small, amounting to just 0.85 percent of global agricultural land. By comparison, genetically modified crops are grown on 2 percent of agricultural land worldwide.

Global meat production increased by 2.6 percent in 2010 to 290.6 million tons. Total production has tripled since the 1970s. Industrial livestock methods behind this increase impose a broad range of environmental and health costs. Per capita meat consumption in the industrial world has plateaued since the late 1980s at around 80 kilograms per year. It remains much higher than the average of 32 kilograms in the developing world, even though that figure has doubled from a quarter-century ago.

Global fish catch is another number at an all-time high—145 million tons in 2009 and expected to climb to 147 million tons in 2010. Once a minor contributor to total fish harvest, aquaculture (fish farming) has increased some 50-fold since the 1950s. It now accounts for 40 percent of total catch and is set to surpass the yield from wild fisheries—which increasingly are troubled by overfishing—within a few years. Yet aquaculture tends to be resource-intensive and often generates large amounts of waste.

On the surface, rising food production looks like good news in a world where hundreds of millions of people go hungry. But it is unequal access to food (and buying power) that lies more directly than insufficient production behind hunger. Meanwhile, the number of overweight people age 15 or older worldwide jumped from 1.45 billion in 2002 to 1.93 billion in 2010, an increase of 25 percent. The share of the population who are overfed thus rose from less than a quarter to 38 percent.

The world's forests shrank by 1.3 percent or 520,000 square kilometers from 2000 to 2010—an area roughly the size of France. A sliver of good news is that the annual rate of net forest loss has decreased from 0.20 percent in the 1990s to 0.13 percent in the first decade of this century.

Payments for ecosystem services are increasingly seen as a way to protect those services—meaning everything from crop pollination to water filtration. Payment schemes for watershed and biodiversity services are currently the primary markets for ecosystem services, estimated to have a combined global value of at least $11 billion in 2008. Smaller markets exist for forest carbon sequestration programs and water quality trading.

Another market-based response to growing environmental problems—carbon trading—is now in use in over 30 countries, but the European Union accounts for three quarters of all such transactions. Between 2008 and 2009, the global volume of transactions increased 80 percent. But because carbon prices plummeted—potentially an indication that all is not well with emissions trading systems as designed—the total value of such transactions increased only 6 percent, thus weakening the signal effect of carbon pricing.

Market-based mechanisms are important but cannot ensure sustainability on their own. Far-sighted public policy and cultural and behavioral changes are essential—and mutually reinforcing. Much of that will happen on the domestic national level. But in a year when the world gathers once more in Rio for a big Earth Summit, it is also clear that international coordination and cooperation are equally critical to turn around environmental and social trends.

Robert Engelman
President

Michael Renner
Project Director

Energy Trends

Thin-film solar panels being installed on the roof of a Walmart store in California

For additional energy trends, go to vitalsigns.worldwatch.org.

Walmart Stores

Oil Market Resumes Growth
after Stumble in 2009

Saya Kitasei and Natalie Narotzky

After falling 1.5 percent between 2008 and 2009 due to the global financial crisis, global oil consumption recovered by 3.1 percent in 2010 to reach an all-time high of 87.4 million barrels per day.[1] (See Figure 1.) About one third of this growth came from China, which now uses over 10 percent of the world's oil.[2] The United States, Brazil, Russia, and the Middle East accounted for an additional 48 percent of the increase.[3] Meanwhile, consumption in the European Union decreased for the fourth consecutive year, falling 1.1 percent.[4] The gap in oil consumption between countries in the Organisation for Economic Co-operation and Development (OECD) and non-OECD countries narrowed, with the two groups respectively accounting for 52.5 and 47.4 percent of total oil consumption in 2010.[5]

In 2010, oil remained the largest source of primary energy use worldwide, but its share of this use fell for the eleventh consecutive year, to 37 percent.[6] Responding to this falling demand, global oil production fell 2.1 percent to 80.3 million barrels per day in 2009.[7] (See Figure 2.) Faced with lowered demand and prices, mem-

bers of the Organization of Petroleum Exporting Countries (OPEC) decided in December 2008 to reduce their production targets by about 4.2 million barrels per day.[8] As a result, OPEC oil production dropped 6.6 percent between 2008 and 2009, its largest annual production decline since 1983.[9]

OPEC and non-OPEC countries (excluding the former Soviet Union) each accounted for almost 42 percent of global oil production in 2010, with the former Soviet Union responsible for 16.8 percent, up from 10.7 percent in 2000.[10] (See Figure 3.) Russia has taken the top producing spot from Saudi Arabia in the last two years.[11]

Although the large share of oil produced in politically unstable regions remains a potential source of oil market volatility, other recent events have created an additional source of risk for global oil markets. On April 20, 2010, a well blowout and fire on the Deepwater Horizon offshore drilling rig in the Gulf of Mexico left oil leaking for months from the partially constructed offshore well.[12] The blowout and oil spill caused public outrage around the world and led a number of governments to put a moratorium on offshore oil drilling until the causes of the accident could be discovered.[13]

The subsequent presidential National Commission on the BP Deepwater Horizon Oil Spill and Offshore Drilling found that "changes in safety and environmental practices, safety training, drilling technology, containment and clean-up technology, preparedness, corporate culture, and management behavior will be required if deepwater energy operations are to be pursued in the Gulf—or elsewhere."[14] The Commission also recommended strengthened safety and environmental regulations on offshore drilling, and the International Energy Agency (IEA) fore-

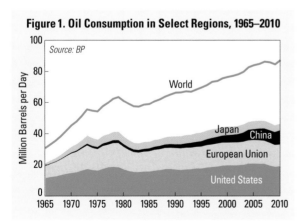

Figure 1. Oil Consumption in Select Regions, 1965–2010

Source: BP

casts that the new regulations that have been called for around the world could mean that projected oil supply from deepwater projects in 2015 could be between 100,000 and 800,000 barrels per day less than previously projected.[15]

Despite the Deepwater Horizon accident, global production increased in 2010 along with prices, rising 2.2 percent to about 82.1 million barrels per day.[16] Nevertheless, the 2010 increase in production was small compared with post-recession growth in coal and natural gas production, which rose 6.3 and 7.3 percent respectively between 2009 and 2010.[17] While the surge in cheap natural gas supplies since 2008 displaced oil in certain markets around the world, it also resulted in the addition of relatively inexpensive natural gas liquids (NGLs— liquid hydrocarbons produced in natural gas wells) to global oil supplies, which are not limited by OPEC production targets and have helped reduce oil prices.

Since the end of 2010, political unrest in the Middle East and North Africa has constrained production in the world's biggest oil-producing region.[18] According to estimates from the International Energy Agency, by the end of May 2011 civil war in Libya—which accounted for 2 percent of the world's oil production in 2010—had removed 132 million barrels of light, sweet crude oil (1.5 million barrels per day) from the world supply.[19]

Concern over the potentially disruptive impact of the Libyan civil war on global oil markets and a lack of stabilizing action by OPEC led the IEA in late June 2011 to release 60 million barrels from its member nations' emergency petroleum reserves for only the third time in history.[20] Of this total, 30 million barrels will come from the United States, 20 million from the European Union's member countries, and 10 million from Asia.[21] This move was intended to complement Saudi Arabia's stated plan to increase its own production, potentially to as much as 10 million barrels per day or more by July 2011, although the IEA and Saudi Arabian actions both drew sharp criticism from other OPEC members, particularly Iran, that are con-

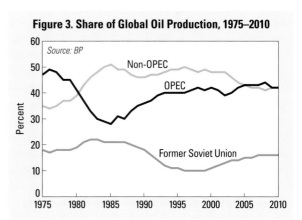

Figure 2. Oil Production in Select Regions, 1965–2010

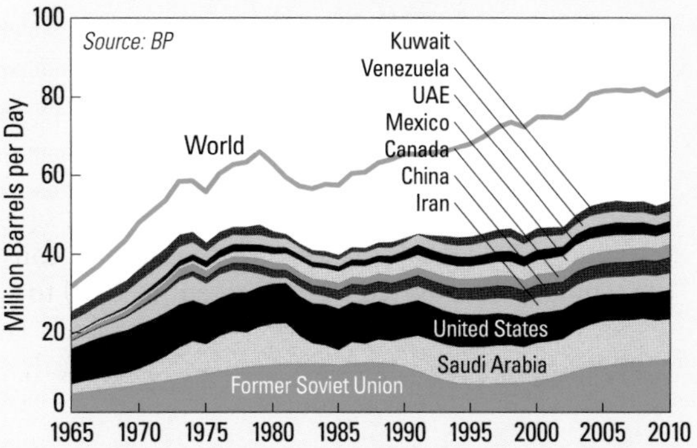

Figure 3. Share of Global Oil Production, 1975–2010

cerned about the dampening effect they would have on oil prices.[22]

After reaching all-time highs in mid-2008, oil prices fell sharply as the global financial crisis drove demand down.[23] (See Figure 4.) With OPEC's decision to cut production targets in the first quarter of 2009, world crude prices began to recover, and average annual prices for West Texas Intermediate crude reached $79.48 per barrel in 2010, which was 25 percent lower than the average 2008 price of $99.69 per barrel.[24]

Against the backdrop of rising oil prices and concerns about supply risk, many countries are paying more attention to their dependence on imports and the stability of the countries from

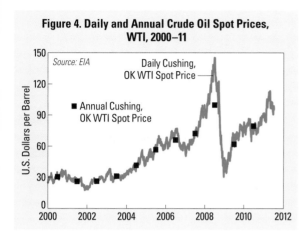

Figure 4. Daily and Annual Crude Oil Spot Prices, WTI, 2000–11

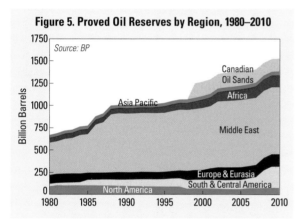

Figure 5. Proved Oil Reserves by Region, 1980–2010

total, exported 7.2 percent more oil in 2010 than in the preceding year.[28] The Asia Pacific region experienced a 10.6 percent increase in exports, with 11.6 percent of total exports coming from that region.[29]

Meanwhile, global proved oil reserves, including natural gas condensate and NGLs in addition to crude oil, have been increasing since 1980, reaching an estimated 1,383 billion barrels in 2010 (1,526 billion including Canadian oil sands).[30] (See Figure 5.) The largest stocks of proved reserves are in the Middle East, which holds 54.4 percent of the world's total.[31]

Oil sands, as well as other unconventional oil sources such as Venezuelan extra-heavy oil, coal-to-liquids, gas-to-liquids, and oil shales, represent huge resources, but their relatively high production and environmental costs will likely prove to be important limiting factors on production.[32]

Oil sands in the Canadian province of Alberta accounted for an additional 143 billion barrels of proved reserves in 2010, equivalent to slightly more than Europe and Eurasia's reserves combined.[33] Canadian oil sands now contribute around half of that country's crude oil production and are expected to provide a growing share, but they are energy- and water-intensive to develop and, in the case of pit mining, can lead to extensive landscape alteration and large waste streams of toxic mining tailings.[34] These environmental concerns have led to strong opposition to oil sands development among certain interest groups in major U.S. and European markets.

Saya Kitasei was a MAP Sustainable Energy Fellow and Natalie Narotzky was a Climate and Energy research intern at Worldwatch Institute. The authors would like to thank Michael Ratner of the Congressional Research Service for his invaluable suggestions.

which they purchase oil. In 2010, the United States imported 50 percent of the oil it needed, compared with 67 percent in Europe.[25] In China, imported oil accounted for 59 percent of total oil consumption, while Japan imported 96 percent of its oil.[26]

The Middle East remains the largest exporter of oil, with 35.3 percent of the world's gross exports in 2010.[27] The former Soviet Union, the second largest exporter, with 16 percent of the

Global Natural Gas Consumption Regains Momentum

Saya Kitasei and Ayodeji Adebola

Global natural gas consumption rebounded 7.4 percent in 2010 after a slight dip in 2009, reaching a record 111.9 trillion cubic feet (Tcf).[1] (See Figure 1.) Strong growth in all regions reflected the end of the recession's dampening effect on energy consumption. In 2010, natural gas accounted for 23.8 percent of global primary energy use, a slight increase over 2009.[2]

Responding to revived demand, global natural gas production increased almost as much as consumption—7.3 percent—to 112.8 Tcf in 2010.[3] Global proved natural gas reserves increased by only 0.3 percent to 187.1 Tcf, or 59 years of current production levels, due in large part to the rapid growth in natural gas production.[4] Most of these reserves are concentrated in the Middle East (40.5 percent) and the former Soviet Union (31.3 percent).[5]

Several organizations have published research in the past few years suggesting that there could be as much global recoverable natural gas resources in unconventional formations as there are in conventional ones.[6] These unconventional resources, including shale gas, coalbed methane, and tight sands, appear to be distributed much more broadly than conventional resources, with a 2011 shale gas assessment by the U.S. Energy Information Administration identifying world-class shale resources in almost every continent studied.[7] The United States and Canada are the only two countries where unconventional gas made up a significant portion of natural gas production in 2010, but Australia, Poland, Germany, the United Kingdom, and China are actively pursuing shale gas development within their own borders.[8]

A number of factors bolstered natural gas markets in North America, including sustained low prices. The world's largest incremental increase in natural gas consumption occurred in the United

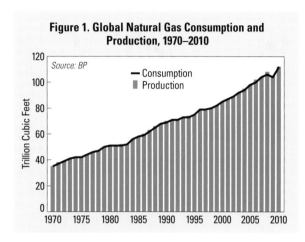

Figure 1. Global Natural Gas Consumption and Production, 1970–2010

Source: BP

States, where consumption leapt by 1.3 Tcf to reach 24.1 Tcf in 2010, which was just over one fifth of global natural gas consumption.[9] The United States also maintained its position as the world's largest natural gas producer for the second year in a row, accounting for just under one fifth of global natural gas production.[10]

The former Soviet Union, after experiencing the largest regional decline in natural gas consumption in 2009, saw a strong rebound in demand in 2010, up 6.8 percent to 21.1 Tcf.[11] Russia, the world's second largest natural gas consumer, accounted for just under 70 percent of this use.[12] Russian production also jumped 11.6 percent, to 20.8 Tcf, just behind the United States, though still not up to its peak levels of 2008.[13] Russia holds almost one quarter of the world's proved natural gas reserves.[14] Production in Turkmenistan, holder of the world's fourth largest natural gas reserves, also recovered 16.4 percent to 1.5 Tcf after an explosion along an export pipeline to Russia halved production in 2009.[15]

European consumption rose as well in 2010, by 7.4 percent to 17.7 Tcf.[16] Production in Norway, Europe's largest producer, reached record levels of 3.8 Tcf, but Europe's proved reserves fell for the seventh consecutive year.[17] Although Europe has accounted for close to one fifth of global natural gas consumption for the last three decades, its share has recently declined, and European usage was overtaken by Asia in 2009.[18]

Asian demand continued to grow rapidly, driven largely by China, which surpassed Japan in 2009 to become Asia's largest natural gas consumer; it used 3.9 Tcf, or 3.4 percent of global natural gas demand, in 2010.[19] China, India, South Korea, and Taiwan all experienced demand growth that year of more than 20 percent.[20] Production in the Asia Pacific region (including Australia and New Zealand) rose by 10.5 percent, led by China, Indonesia, and Malaysia, and contributed 15.4 percent to global natural gas supplies.[21] China's 12th Five-Year Plan, which covers 2011–15, includes the goal of increasing natural gas's share in the country's primary energy mix to 8.3 percent by 2015—about double its current share—sending a strong signal that Chinese natural gas consumption will continue to increase rapidly over the next five years.[22]

Although the Middle East is home to some of the richest natural gas resources in the world, with Iran and Qatar alone accounting for 29.4 percent of global proved reserves as of 2010, the region has only recently begun to develop the infrastructure needed for domestic consumption.[23] Consequently, despite a 13.2 percent increase in the region's natural gas production, driven mainly by Qatar, usage in the Middle East grew by only 6.2 percent, with the remainder going to exports.[24] The Middle East is widely regarded as one of the world's largest potential growth markets.[25]

Both Africa and South America have comparatively immature natural gas markets due to relatively modest production and a lack of robust transmission and distribution infrastructure, with consumption at just 3.3 and 4.7 percent,

respectively, of the global total in 2010.[26] Nevertheless, both regions exhibited growth in both consumption and production. Nigeria, where flaring of natural gas associated with oil production has long been criticized as a major source of greenhouse gas emissions and other environmental damage, has improved its capture rate of natural gas, and production there rose 35.7 percent over 2009.[27]

Gas flaring continues to be a challenging environmental and energy issue throughout the world. It is estimated that 5–5.5 Tcf of associated gas, or the equivalent of about 5 percent of global natural gas production, is flared annually.[28] Satellite data indicate that Russia and Nigeria were the two largest gas-flaring countries in the world, with an estimated 1.4 Tcf and 0.5 Tcf of gas flared, respectively, in 2009.[29] Global gas flaring fell from 5.2 Tcf in 2009 to 4.7 Tcf in 2010, marking the fifth consecutive year that gas flaring has dropped worldwide despite rising oil production: in fact, between 2005 and 2010 gas flaring decreased 22 percent.[30] Russia and Nigeria have achieved significant reductions in flaring through the use of natural gas-fired power plants to displace diesel generators.[31] Kazakhstan has also been credited with cutting its gas flaring activities by a third in five years.[32]

Reenergized global gas demand drove average prices up from their 2009 lows in all markets but the European Union. In the United States, natural gas was traded on the Henry Hub spot market at an average of $4.39/million Btu (MMBtu) in 2010, a 13 percent increase over 2009 levels.[33] (See Figure 2.) Over the same period, average spot prices on the British National Balancing Point (NBP) Hub rose 35.2 percent to $6.56/MMBtu.[34] Prices in Asia, the market that saw the largest rate of increase in consumption, remained the highest, with the average price of liquefied natural gas (LNG) reaching $10.91/MMBtu in 2010.[35] Yet even these high prices were lower than global oil prices in 2010, which reached $13.47/MMBtu.[36]

The sharp divergence between oil and natural gas price trends in 2009 and 2010 put downward pressure on oil-indexed prices in long-term contracts, particularly in Europe—where the coincidence of falling demand, the availability of cheap LNG originally intended for U.S. markets, and the emergence of spot markets has threatened the dominance of oil indexation in long-term contracts with major suppliers such as Russia's Gazprom.[37] Average natural gas prices in the European Union fell 6 percent to $8.01/MMBtu in 2010.[38]

The share of global natural gas trade represented by LNG passed 30 percent in 2010 for the first time on record.[39] (See Figure 3.) The total volume of LNG traded globally reached 10.5 Tcf in 2010, up 22.6 percent from 2009.[40] About half of the increase in exports came from Qatar, whose LNG exports rose 53.2 percent to 2.67 Tcf (about one quarter of LNG traded in 2010).[41] Exports from Australia, Indonesia, and Malaysia, which together represent an additional 29.3 percent of exports, increased 9.4 percent.[42]

Global liquefaction (LNG export) capacity reached 270.9 million tons per annum (MMtpa) by the end of 2010, which was 58 percent greater than in 2005.[43] The recent rapid build-out of liquefaction capacity was driven by Qatar, which is now home to 26 percent of global liquefaction capacity.[44] (See Figure 4.) An additional 57.1 MMtpa of liquefaction is under construction, 63 percent of which is located in Australia, where conventional and coalbed methane resources are expected to become a major source of global natural gas exports in the near term.[45]

Asia continued to be the largest consumer of LNG, with Japan alone importing 3.30 Tcf in 2010—an 8.8 percent increase over 2009 and 31.4 percent of all LNG traded.[46] South Korea imported 1.57 Tcf, up 29.4 percent.[47] And Chinese LNG imports, while more modest at 0.45 Tcf, grew 67.8 percent since 2009.[48] European imports reached 3.10 Tcf in 2010.[49] Spain maintained its lead as the largest European LNG importer at 0.97 Tcf, but the largest growth in LNG imports occurred in Italy (213.1 percent) and the United Kingdom (82.3 percent).[50] U.S. LNG imports fell for the third year in a row, reflecting the availability of abundant, low-cost natural

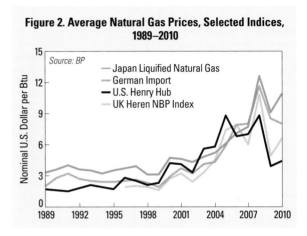

Figure 2. Average Natural Gas Prices, Selected Indices, 1989–2010

Source: BP

— Japan Liquified Natural Gas
— German Import
— U.S. Henry Hub
— UK Heren NBP Index

Figure 3. Volumes of LNG and Pipeline Gas Traded, 2001–10

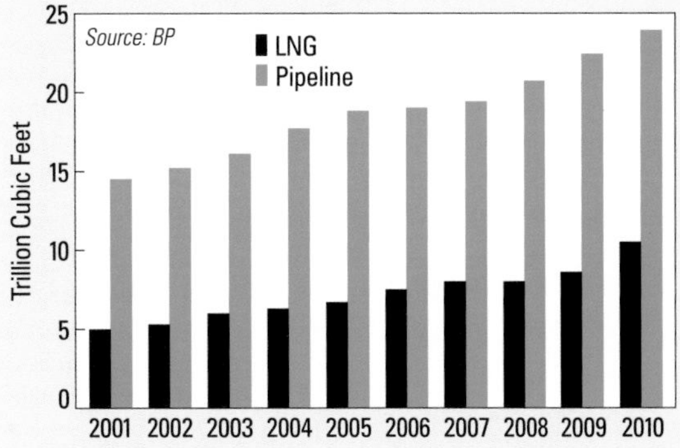

Source: BP

■ LNG
■ Pipeline

gas supplies in the North American market.[51]

Although East Asia, especially Japan, South Korea, and Taiwan, have traditionally dominated global LNG imports, the geography of the global LNG market is undergoing a transformation. As of 2010, some 23 countries had LNG import capabilities, with global regasification capacity at 572 MMtpa.[52] Almost one third of this capacity is located in Japan, which had 28 import terminals in operation at the end of 2010.[53] Of the six countries that began importing LNG between 2005 and 2010, five are in South America or the Middle East (the sixth is Canada).[54]

Figure 4. Share of Global Liquefaction Capacity by Country,

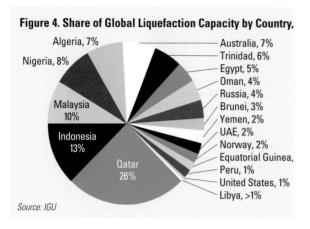

Source: IGU

Natural gas trade via pipelines increased by a more modest 6.9 percent to 23.92 Tcf in 2010.[55] Russia remained the world's largest exporter of natural gas, with pipeline exports alone amounting to 6.58 Tcf, or 27.5 percent of global pipeline trade.[56] Nevertheless, Russia's position as the main supplier for much of Eastern Europe and the sole buyer of natural gas produced in the Caspian region has begun to weaken: in 2010, a new pipeline opened to transport natural gas from Turkmenistan to China, which purchased 0.13 Tcf from the Central Asian country in the pipeline's inaugural year of operation.[57] Turkmenistan also constructed a short pipeline to increase its exports to Iran.[58]

On the other shore of the Caspian, by the end of 2011 Azerbaijan's Shah Deniz Consortium is due to select among four competing proposals a pipeline that would carry to Europe natural gas from its Shah Deniz field (providing 350 billion cubic feet per year) without going through Russia.[59] The European Union's search for supply alternatives to pipeline imports from Russia has intensified since a pricing dispute between Russia and Ukraine left much of Eastern Europe without heat for three weeks in January 2009.[60]

Two additional developments in 2011 significantly affected the trajectory of global natural gas markets. First, political unrest in a number of natural gas–producing countries in North Africa took some production offline during 2011. Libyan exports to Italy, which had hovered around 900 million cubic feet per day, were shut off at the end of February and did not resume until mid-October.[61] And attacks on a facility that began after the overthrow of former Egyptian President Hosni Mubarak curtailed shipments of natural gas through a pipeline that serves Jordan and Israel between July and October 2011.[62]

Second, an earthquake and subsequent meltdown at the Fukushima Daiichi nuclear power plant in March 2011 has led Japan and other countries to reevaluate their dependence on nuclear energy for power generation.[63] Natural gas is likely to play a major role in filling the gap left by idled and phased out nuclear plants. Japan's power utilities purchased record volumes of LNG to cover the shortfall in their nuclear generation after the disaster at Fukushima, pushing Japan's LNG prices to $16.372/MMBtu in August 2011.[64] The unanticipated spike in public opposition to nuclear power can only increase global natural gas demand in the coming decade.

Saya Kitasei was a MAP Sustainable Energy Fellow and Ayodeji Adebola was a Climate and Energy research intern at Worldwatch Institute.

Nuclear Generation Capacity Falls

Matt Lucky

Global installed nuclear generating capacity declined in 2011, falling to 366.5 gigawatts (GW).[1] (See Figure 1.) As of October 2011 there were 433 nuclear reactors in operation around the world, compared with 441 at the beginning of the year.[2] Just over 5.1 GW of installed capacity has been added since the beginning of 2010, with new connections to the grid in China, India, Iran, Pakistan, Russia, and South Korea.[3] Over that same period, nearly 11.5 GW of installed capacity has been shut down in France, Germany, Japan, and the United Kingdom.[4] More than 8.4 GW of installed capacity was shut down in Germany alone in 2011.[5]

In 2010 nuclear construction starts worldwide reached their highest levels since 1980, with 16 new reactors beginning construction at 15 different locations.[6] Construction starts decreased significantly in 2011, however, with only India and Pakistan starting to build one new reactor each during this time.[7] (See Figure 2.) There are 65 reactors currently under construction in 14 countries around the world.[8] When these are completed, they will provide 62.6 GW of additional installed capacity.[9] Twelve of these 65 reactors, however, have been "under construction" for more than 20 years.[10]

The construction of Flamanville 3, the first nuclear power plant to be built in France in 15 years, has been delayed for economic and structural engineering reasons.[11] It was originally supposed to come online in 2012 at a cost of 3.3 billion euros, but it is now expected to start up in 2016 at a cost of 6 billion euros.[12]

China has 27 reactors with a total capacity of 27.2 GW currently under construction.[13] In fact, China accounted for 10 of the 16 reactor construction starts in 2010.[14] The government initiated 9.9 GW of construction in 2010, accounting for 62 percent of global construction

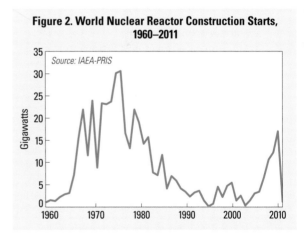

Figure 1. World Electrical Generating Capacity of Nuclear Power Plants, 1960–2011

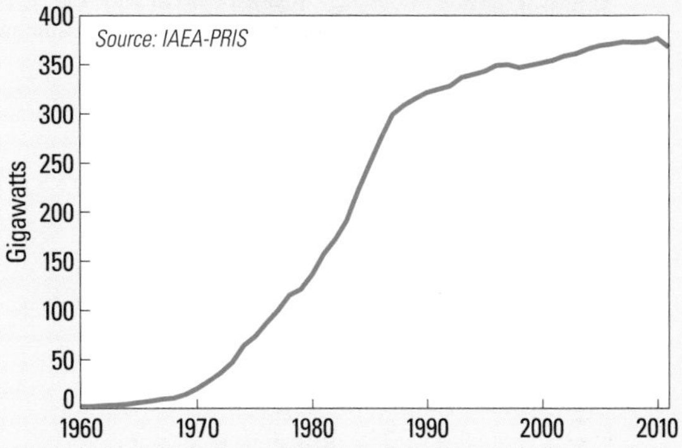

Source: IAEA-PRIS

Figure 2. World Nuclear Reactor Construction Starts, 1960–2011

Source: IAEA-PRIS

started that year in terms of rated capacity.[15] (See Figure 3.)

Construction starts steadily increased after 2004 as China began to pursue nuclear energy more aggressively. Recently connected reactors in China and India alone account for 56 percent of new grid-connected nuclear capacity since

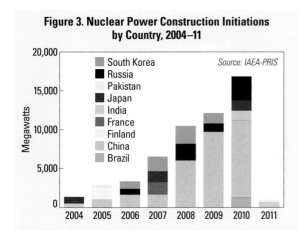

Figure 3. Nuclear Power Construction Initiations by Country, 2004–11

Source: IAEA-PRIS

Legend: South Korea, Russia, Pakistan, Japan, India, France, Finland, China, Brazil

2006.[16] (Only Iran, Japan, Pakistan, Romania, Russia, and South Korea have also added nuclear capacity since 2006.)[17] Reactor construction starts fell dramatically in 2011, however, as countries like China temporarily postponed construction starts in reaction to the meltdown and shutting of reactors at the Fukushima Daiichi plant in Japan following the March 2011 earthquake and tsunami.

In the first 10 months of 2011, a total of 13 nuclear reactors were permanently shut down.[18] The total number of decommissioned nuclear reactors around the world is now 138, with a cumulative total capacity of more than 49.1 GW.[19] (See Figure 4.) The average age of all decommissioned reactors has risen slightly to 23 years, as the reactors shut down in Germany had an average life of nearly 33 years.[20]

In 2009, the U.S. Nuclear Regulatory Commission received 26 nuclear reactor permit applications, yet construction to date is planned for only 4 of these facilities.[21] This is mostly a consequence of high costs, slowed electricity demand resulting from the economic slowdown, and lower natural gas prices.[22] There is only one nuclear power plant currently under construction in the United States: the Watts Bar 2 reactor in Tennessee. Construction there began in 1973, but it was postponed in the 1980s and only started up again in 2007.[23]

In 2010, the Obama administration set aside $8.3 billion for loan guarantees for the construction of two nuclear reactors in Georgia.[24] As part of his budget proposal in February 2011, the president announced the allocation of an additional $36 billion for loan guarantees for future construction of nuclear reactors.[25] Following the disaster at Fukushima, Energy Secretary Steven Chu reaffirmed the administration's commitment to nuclear development in the future, but he emphasized that the United States will need to learn from the mistakes leading up to the events at Fukushima.[26]

As noted earlier, interest in nuclear energy has been slowed by the events at Fukushima. China suspended all approvals for new nuclear reactors on March 16, 2011.[27] It also froze construction of 25 reactors and increased the number of nuclear safety staff from 300 to 1,000.[28] It is expected, however, that China's approval process for reactor applications will resume in mid-2012.[29] Overall, the likelihood of China significantly reducing its aggressive growth in nuclear generation remains low as the country intends to meet rapidly growing energy demand and ambitious carbon dioxide reduction targets.[30]

Nearly seven months after the events at Fukushima, only 10 of Japan's 54 nuclear reactors are connected to the grid.[31] Many of these have passed safety maintenance inspections, but the government has been slow to reintroduce them to the grid. Of the six reactors at Fukushima, four will remain permanently decommissioned.[32]

Germany has decided to phase out its nuclear program, which at the beginning of 2011 contributed 20 percent of the country's electricity production.[33] In addition to the eight reactors shut down following Fukushima—which accounted for roughly 41 percent of the country's nuclear generating capacity—Germany plans to take its nine remaining nuclear reactors offline by 2022.[34]

The Swiss government approved a phaseout plan for its nuclear plants following Fukushima as well. The first of its five reactors will be taken offline in 2019 and the final one will be decommissioned by 2034.[35] These five will not be replaced by a new generation of nuclear reactors in the future, as the approval process for reactors has been suspended.[36] Italy and Venezuela have also indefinitely postponed nuclear reactor planning and construction after Fukushima.[37]

Yet nuclear energy remains an important source of electricity for many countries. France stated it will continue to use nuclear power as its main source of energy.[38] In the week following Fukushima, Russia agreed to finance a two-reactor nuclear power plant in Belarus for $6 billion.[39] And in October 2011, Belarus signed a contract to begin construction of the nation's first nuclear reactors in 2012.[40]

Nuclear power's share of the world's commercial primary energy fell to 5.2 percent in 2010, compared with a peak of 6.4 percent in 2001 and 2002.[41] (See Figure 5.) Nuclear power supplied 13.0 percent of the world's electricity in 2010, down from 16.8 percent in 2000.[42] Only four countries—the Czech Republic, Romania, Slovakia, and the United Kingdom—increased their share of electricity generation from nuclear power by greater than 1 percentage point from 2009 to 2010.[43]

Nuclear power's share of global electricity production is likely to continue to decline in the future. Even with recent construction starts worldwide, the world would need to commence construction and bring online an additional 18 GW by 2015 just to maintain current generation levels.[44] And as the average age of operating nuclear power plants is already 26 years, the world would need to bring online another 175 GW of nuclear power to make up for the decommissioning of plants between 2015 and 2025.[45] In the aftermath of Fukushima and in the context of the global financial crisis, achiev-

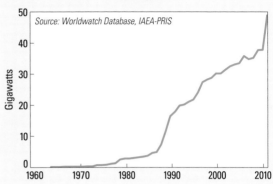

Figure 4. Nuclear Capacity of Decommissioned Plants, 1964–2011

Source: Worldwatch Database, IAEA-PRIS

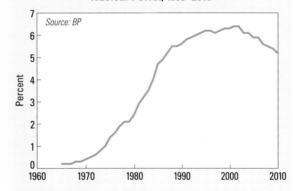

Figure 5. Share of World Primary Energy from Nuclear Power, 1965–2010

Source: BP

ing these amounts of new nuclear capacity seems unlikely.

Matt Lucky is a MAP Sustainable Energy Fellow at Worldwatch Institute.

Global Wind Power Growth Takes a Breather in 2010

Mark Konold

Global wind power capacity increased by 38,000 megawatts in 2010 to a total of 197,000 megawatts.[1] (See Figure 1.) The global market grew by 24 percent, down from a 31 percent increase in 2009.[2] But by now, installed capacity is three times greater than it was in 2006 and nine times what it was a decade ago.[3] The European Union had 43 percent of total installed capacity; in terms of individual nations, China and the United States lead the field at 23 and 20 percent, respectively.[4] (See Figure 2.)

China overtook the United States in terms of total installed wind capacity in 2010, bringing its total to just under 45,000 megawatts. This is an increase of more than 70 percent from 2009, meaning the country maintained its leading position in terms of capacity additions.[5] (See Figure 3.) In 2010 it connected just under 14,000 megawatts to the country's grids, bringing total grid connectivity to about 31,000 megawatts.[6] This means that 14,000 megawatts, or 31 percent of the country's turbine potential, sits idle. China's National Energy Bureau

expects to bring total grid connectivity to 55,000 megawatts this year.[7] The country continues to experience difficulties with long-distance transmission and the ability of some of its grids to absorb electricity from wind sources.[8] According to analysts, this issue should abate in time with smart grid investments, but these problems are likely to persist in the near term.[9] The State Grid Corporation of China has said that it plans to connect 90,000 megawatts of capacity to the grid by 2015.[10] Of the country's major wind power bases, Gansu Jiuquan province ended 2010 with about 5,000 megawatts installed, followed by Inner Mongolia East and Heibei at 4,211 and 4,160 megawatts, respectively.[11] China's next five-year plan, which begins in 2011, calls for an increase of 70 gigawatts by 2015.[12]

In 2010, the United States saw the slowest rate of growth in almost a decade, but it maintained its second-place position in the world in terms of installed capacity. The country ended the year with 40,180 megawatts of installed capacity, up from 35,159 megawatts in 2009.[13] A lack of long-term predictable federal policies continues to fuel a boom-bust cycle in the United States.[14] Texas remained the leading state, with 10,085 megawatts of installed capacity—up from 2009's total of 9,403 megawatts.[15] Iowa and California finished a distant second and third in 2010, with 3,675 and 3,177 megawatts, respectively.[16] Wind met approximately 2 percent of U.S. electricity needs.[17]

The European Union saw a 10-percent slowdown in its rate of wind growth compared with 2009.[18] Germany still managed to add 1,493 megawatts to its portfolio to reach a total installed capacity of 27,214 megawatts, the most in the European Union.[19] But top honors for installation in 2010 went to Spain, with 1,516

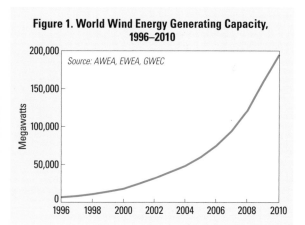

Figure 1. World Wind Energy Generating Capacity, 1996–2010

Source: AWEA, EWEA, GWEC

megawatts of added capacity.[20] France added 1,086 megawatts, and the United Kingdom added 962 megawatts.[21] Also of note was Romania's jump from 14 megawatts in 2009 to 462 megawatts by the end of 2010.[22]

India added 2,139 megawatts, bringing its total capacity to 13,065 megawatts, a jump of roughly 19.5 percent from 2009.[23] The state of Tamil Nadu has the most installed capacity on the subcontinent, with 4,907 megawatts, followed by Maharashtra with 2,078 megawatts.[24] Progress may stall, however, as the country seeks to overhaul its tax code, a move that may eliminate accelerated depreciation of equipment. This incentive has been a driving force behind wind's growth on the subcontinent.[25] Japan installed an additional 10 percent of wind capacity in 2010, ending the year at 2,304 megawatts installed.[26]

Latin America continued to hold a rather small share of global installed capacity—2,008 megawatts, or 1 percent of the global market.[27] Brazil and Mexico continued to be regional leaders, with 931 megawatts and 519 megawatts, respectively.[28]

It was a good year for offshore wind capacity. In 2010, Europe alone saw the successful connection of 308 turbines to the grid, raising its total offshore capacity to just under 3,000 megawatts.[29] An additional 1,000–1,500 megawatts is expected to come online during 2011.[30] In the second half of 2010, China's first offshore wind farm, Donghai Bridge, began providing electricity to the grid.[31] The farm consists of 34 turbines and has 102 megawatts of capacity. By 2020 the country should have 30 gigawatts of installed offshore capacity, according to the National Development and Reform Commission.[32] It is projected that China will spend more than $4.2 billion on offshore wind in the next four years.[33]

The United States continues to lag behind Europe and China in this field. After nine years of opposition, the Cape Wind project off the coast of Massachusetts was finally approved.[34] And although Gamesa and shipbuilder Northrop Grumman opened an Offshore Wind Technol-

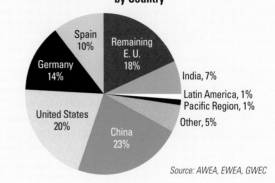

Figure 2. Installed Wind Energy Generating Capacity, by Country

Source: AWEA, EWEA, GWEC

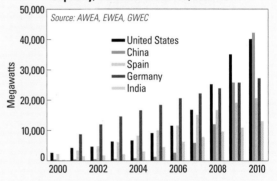

Figure 3. Annual Installed Wind Energy Generating Capacity, Selected Countries, 2000–10

Source: AWEA, EWEA, GWEC

ogy Center in 2010 to explore designs for offshore systems that will be installed for U.S. markets, to date the United States has yet to install a single offshore turbine.[35]

In 2010, wind turbine prices fell below $1.4 million per megawatt, a level that had not been seen since 2005.[36] Increased demand and significant technological improvements helped bring costs 19 percent below the highs seen in 2007.[37] A total of $96 billion was invested in wind energy installations in 2010, a 31 percent increase from 2009.[38] Together, China and large European offshore wind farms accounted for 38 percent of total wind investment in 2010.[39] In terms of

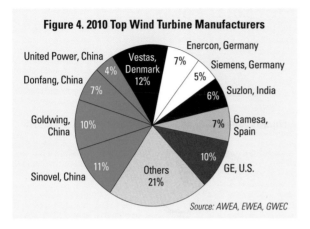

Figure 4. 2010 Top Wind Turbine Manufacturers

Source: AWEA, EWEA, GWEC

overall turbine sales, Vestas remains the world leader, with 12 percent of the world market.[40] In an interesting change, Chinese manufacturer Sinovel moved into the number two spot, the first of four Chinese manufacturers in the top 10. American producer GE had to settle for a tie for third place. (See Figure 4.)

Mark Konold is a project manager for the Climate and Energy Program at Worldwatch Institute.

Another Record Year for Solar Power, but Clouds on the Horizon

Sam Shrank and Matthias Kimmel

The photovoltaic (PV) and electric solar thermal markets had another record year in 2010. With an estimated 16,700 megawatts (MW) of capacity installed, PV continued to dominate the solar landscape.[1] The additions in 2010 accounted for almost 2.5 times more growth than in 2009 and were more than the total existing PV capacity as recently as 2008.[2] With installed capacity at almost 40,000 megawatts by the end of 2010, PV cells produced enough electricity to power about 13 million households.[3] Additionally, about 500 MW of solar thermal electric power plants came online in 2010, bringing the total operating capacity of such plants to roughly 1,100 MW.[4] In Germany and Spain, solar energy provided by PV and thermal electric plants now meets, respectively, about 2 percent and 2.6 percent of electricity demand.[5]

As has been the case for some years, Europe was responsible for the lion's share of world PV growth in 2010, installing over 13,000 MW.[6] (See Figure 1.) For the first time, Europe installed more PV than wind capacity, with Germany and Italy leading the way.[7] Germany continued to be the world's biggest market, with 7,400 MW of new PV capacity connected to the grid in 2010.[8] This accounts for roughly 40 percent of global PV additions and is equivalent to total global capacity additions in 2009.[9] Germany, Europe's largest economy, now has 17,300 MW of PV installed.[10] Italy was again the second largest PV market; the 2,300 MW in new installations there in 2010 was four times more than was added in 2009 and pushed Italy's aggregate official capacity to almost 3,500 MW.[11] There is reason to think that installed capacity is much higher, however. Some 5,248 MW of capacity was listed under the country's feed-in tariff registry as of May 2011, much of which could have been added during 2010.[12]

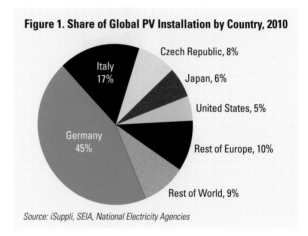

Figure 1. Share of Global PV Installation by Country, 2010

- Germany 45%
- Italy 17%
- Czech Republic, 8%
- Japan, 6%
- United States, 5%
- Rest of Europe, 10%
- Rest of World, 9%

Source: iSuppli, SEIA, National Electricity Agencies

The Czech Republic continued its fast ascent into the group of major PV markets, adding 1,331 MW in a country with little previous capacity.[13] France added 719 MW, surpassing 1 gigawatt (GW) in total installations.[14] Spain, which dominated the PV landscape in 2008 and was long one of the most important solar markets, installed 371 MW in 2010, slightly recovering from the severe crash the market experienced in 2009.[15] Outside Europe, the major markets included Japan and the United States. Japan was the world's fifth largest installer in 2010, with 950 MW added.[16] The United States installed 937 MW and now has 2,100 MW of total capacity.[17]

Solar thermal power is also gaining momentum, with three times as much growth in 2010 as in 2009.[18] Spain was responsible for roughly 60 percent of global concentrated solar power installations, with 350 MW installed in 2010.[19] The seven new facilities, each having a capacity of 50 MW, made up more than 60 percent of Spain's total solar thermal capacity of 582

MW at the end of 2010, which helped the country surpass the United States as the world's largest market.[20]

The United States installed 78 MW of solar thermal power in 2010, compared with just 6 MW in 2009, bringing the country's total existing capacity to 510 MW.[21] The amount added in 2010 was mainly due to the 75 MW Martin Next Generation Solar Energy Center plant in Indiantown, Florida—the second largest solar facility in the world and the first hybrid facility that is connected to a combined-cycle power plant.[22]

In terms of production, 23.9 GW of PV cells were manufactured worldwide in 2010, up from 10.7 GW in 2009.[23] (See Figure 2.) The market share of thin-film cells fell for the first time since 2005, from 17 percent in 2009 to 13 percent in 2010.[24] The 38-percent growth in thin-film capacity in 2010 was far below the 99 percent growth in 2009.[25] First Solar, the industry leader, saw its market share fall to 26 percent, its lowest level since 2006.[26]

Figure 2. Annual Global Production of Photovoltaic Cells, 1980–2010

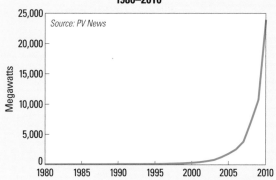

Some 80 percent of all solar modules produced came from Asian manufacturers, with companies based in China and Taiwan contributing 59 percent—an increase of 10 percentage points relative to 2009.[27] Suntech Power and JA Solar, both Chinese firms, were the top cell manufacturers, producing 1.57 GW and 1.46 GW respectively.[28]

Various policies and market mechanisms have been tapped to stimulate solar power demand. So far, feed-in tariffs have been the most prevalent and the most successful. Almost all PV capacity added in Europe since 1997 has been commissioned in areas under feed-in policies that guarantee producers a fixed tariff or premium over a certain period of time for every kilowatt-hour (kWh) they feed into the grid.[29] These laws often incorporate downward-trending payments in order to account for improvements in technology.

But many European governments cut their feed-in tariffs in 2010 or looked to reduce other incentives. (See Figure 3.) Spain amended its policy for the third time in 2010, reducing payments by 5 and 25 percent, respectively, for small and large rooftop PV installations and by 45 percent for ground-mounted installations.[30] In the Czech Republic, a retroactive tax on photovoltaic installations with a capacity of more than 30 kilowatts came into force on January 1, 2011, taxing all installations connected to the grid in 2009 and 2010 at 26 percent over the next three years.[31]

Both Germany and Italy introduced flexible price mechanisms in the first half of 2011 that will cut their PV feed-in tariffs, depending on previous capacity installations. In Germany, for instance, the new scheme could lead to tariff reductions of between 1.5 and 24 percent by the beginning of 2012.[32] These policies aim to avoid excessive capacity additions and have tariffs adapt faster to changes in the PV market.

Part of the motivation for cutting incentives comes from a significant drop in the cost of manufacturing PV panels. Costs fell in 2010 as producers saw significant drops on several important manufacturing costs, made improvements on technical parameters such as efficiency and yield, and ramped up production.[33] Component prices dropped at a rate that for the first time exceeded subsidy cuts.[34] For example, tier 1 Chinese crystalline silicon (C-Si) module prices declined from $2.40 per watt on average in 2009 to $1.75 per watt in 2010.[35]

Increasing commoditization of C-Si technology allowed high commercial development of the technology in low-cost regions such as China and Taiwan.[36] Low-cost capital available from Chinese state banks allowed firms in China and Taiwan to scale up manufacturing capacity at much lower costs than their peers could arrange.[37] In 2010, European firms began to use Chinese module producers to supply them with panels that they could then sell under their own high-profile brand names.[38]

The decreased market share for thin-film technologies can be attributed to falling PV costs and a commensurate increase in the production of C-Si and other modules.[39] Preliminary data for 2010 show installed costs declining in California and New Jersey, the two largest markets in the United States. In California, average installed costs through the Solar Initiative Program were $6.40 per watt during the first 10 months of 2010, $1 below the 2009 average.[40] In New Jersey, average costs through June 2010 were $5.90 per watt, which was $1.20 lower than in 2009.[41]

With growth expected to slow in the leading European markets due to reduced incentives, particularly in Germany, the solar industry is looking to new markets. Many have high hopes for India, with its national solar mission and large unmet demand in rural areas. Others eye the United States, growing rapidly and with attractive incentives and good resources in many states. But neither of these markets is poised for an immediate boom. Growth in China is also expected to be high, but that market will likely be captured almost entirely by domestic manufacturers. Currently less than 5 percent of

Figure 3. Feed-in-Tariff Prices Paid for Rooftop Installations (30–100 kW), Selected Countries

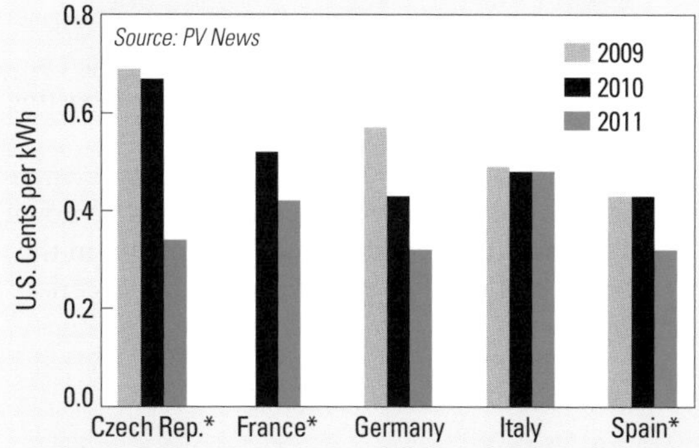

Source: PV News

*Czech Republic: 26% tax on 2009 and 2010 tariff; France: 500 MW cap in 2011; Spain: 488 MW cap in 2011

China's module production is used domestically.[42]

Thus, many market analysts predict a period of significant oversupply of solar modules, particularly PV panels.[43] Costs are likely to come down, as the major producers, who have increased their manufacturing capacity, accept lower profit margins in the face of demand forecasted to grow only modestly.[44] So although 2011 may be a slower year for the solar industry, it may also bring solar power closer to cost competitiveness.

Sam Shrank was a MAP Sustainable Energy Fellow and Matthias Kimmel was a Climate and Energy research intern at Worldwatch Instutite. The authors would like to thank Janet Sawin for her valuable feedback on an earlier draft.

Biofuels Regain Momentum

Sam Shrank and Farhad Farahmand

Global biofuel production increased by 17 percent in 2010 to reach an all-time high of 105 billion liters.[1] (See Figure 1.) The increase exceeded the 10 percent growth experienced in 2009, when production was at 90 billion liters.[2] Biofuels provided 2.7 percent of all global fuel for road transportation—an increase from 2 percent in 2009.[3]

The two biofuel alternatives to fossil fuels for transportation largely consist of ethanol and biodiesel. Ethanol is primarily made by fermenting the sugars in corn and sugarcane, while biodiesel is produced from fats and vegetable oils. The world produced 86 billion liters of ethanol in 2010, which was 18 percent more than in 2009.[4] World biodiesel production rose to 19 billion liters in 2010, a 12 percent increase from 2009.[5]

The United States again led the world in ethanol production in 2010 at 49 billion liters, or 57 percent of world output.[6] Brazil was the second largest producer, at 28 billion liters, which was 33 percent of the world total.[7] Distant followers included China, Canada, France, Germany, and Spain, each producing less than 2.5 percent of world supply.[8] No other countries showed significant changes in ethanol production.

In 2010, ethanol production in the United States grew by 8.4 billion liters, equaling the 2009 growth rate of 20 percent.[9] Corn is the primary feedstock for U.S. ethanol, which supplies 4 percent of the nation's road transport fuel.[10] Ethanol-gas blends for U.S. vehicles that are model year 2001 and newer can be as high as 15 percent ethanol (E15), though in most cases the percentage blended is much lower.[11] Because of unsteady production in Brazil, the United States became a net ethanol exporter for the first time, sending a record 1.3 billion liters abroad, an increase of 300 percent over 2009.[12] The largest markets for U.S. exports were Canada, Jamaica, the Netherlands, the United Arab Emirates, and Brazil.[13]

Sugarcane is the only source of ethanol in Brazil, where production rose by 2 billion liters in 2010, a 7 percent increase.[14] This represented a rebound from the 3 percent drop in production in 2009, even as adverse global weather led to rising sugar prices and was economically disadvantageous for sugarcane ethanol production.[15] Sugarcane ethanol supplies 41.5 percent of the energy (48 percent of the volume) for light-duty transportation fuels in Brazil.[16]

The European Union remained the center of biodiesel production, accounting for 53 percent of global output in 2010.[17] Growth slowed there dramatically, however, falling from 19 percent in 2009 to just 2 percent in 2010.[18] The top biodiesel producers worldwide were Germany (2.9 billion liters, a 12 percent increase) and Brazil (2.3 billion liters, a 46 percent increase and responsible for one third of global growth).[19] (See Figure 2.) Other notable producers include

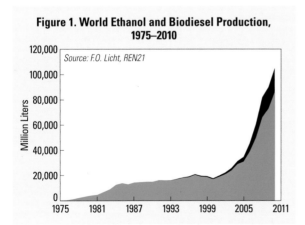

Figure 1. World Ethanol and Biodiesel Production, 1975–2010

Source: F.O. Licht, REN21

Argentina, which increased production by 57 percent to 2.1 billion liters in 2010, and France, whose production fell by 0.6 billion liters to 2.0 billion liters in 2010.[20] Asia produced 12 percent of the world's biodiesel, a 20 percent increase from 2009, mostly from palm oil in Indonesia and Thailand.[21]

The global increase in ethanol output was principally caused by large fuel players in the United States entering the industry in response to high oil prices. Valero, Flint Hills, Sunoco, and Murphy Oil each invested in hundreds of millions of liters of ethanol plant capacity, and Pacific Ethanol returned from the brink of bankruptcy to open four new plants.[22]

Internal dynamics in Brazil also opened the door to U.S. ethanol exports. Unfavorable weather in 2010 drove down the global sugar yield, raised sugar prices, and favored sugar production in Brazil over that of sugarcane ethanol.[23] At the same time, high oil prices led many Brazilians, who drive flex-fuel cars that can run on either fossil fuels or biofuels, to increase their use of ethanol.[24]

The growth of low-cost international competitors from Canada, Argentina, Indonesia, and elsewhere as well as the rising costs of rapeseed oil—the leading biodiesel feedstock in Europe—meant that Europe exploited only 40 percent of its production capacity and relied on imports to meet the rest of its demand.[25] Biodiesel accounts for the majority of biofuels consumed in the region, but many countries may switch from biodiesel to ethanol in the future because of a recent European Commission report on the indirect land use impacts of biodiesel.[26] Ethanol crops have a higher energy content than biodiesel crops, and the report shows that their production may involve lower greenhouse gas emissions.[27] The European Union currently has a 5.75 percent blending mandate for biofuels.[28]

Argentina's biodiesel industry grew in response to a new B7 blending mandate (7 percent biodiesel and 93 percent diesel) as well as to favorable conditions for growing soybeans.[29] Virtually all of the 1.5 billion liters of Argentinean exports, representing 71 percent of total

Figure 2. Biodiesel Production in Million Liters, 2010

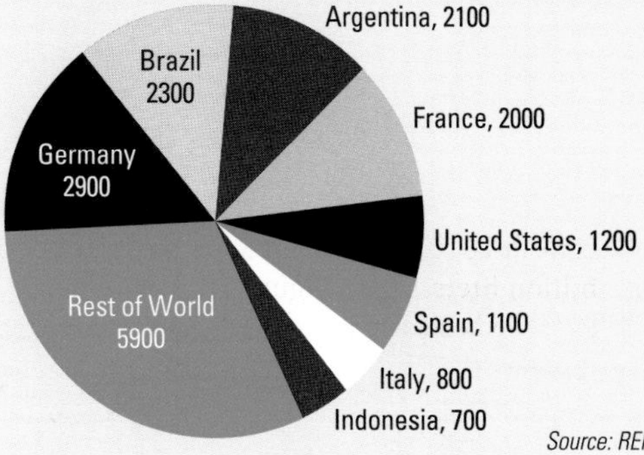

Argentina, 2100
France, 2000
United States, 1200
Spain, 1100
Italy, 800
Indonesia, 700
Brazil 2300
Germany 2900
Rest of World 5900

Source: REN21

production, went to Europe.[30] Argentina has 19 biodiesel production facilities; Brazil, which has a B2 blending mandate that will be increased to B5 in 2013, has 64.[31]Argentinean producers continue to invest in facilities, and a mandate as high as B20 could be instituted in the next four years.[32]

Federally, Canada has E5 (5 percent ethanol and 95 percent gasoline) and B2 mandates, and four Canadian provinces have individual mandates up to E8.5.[33] In China, nine provinces require E10 blends; the country produced 2.1 billion liters of fuel ethanol in 2010 despite lacking a federal blending mandate.[34] The other large biofuel producers in Asia include Indonesia, which has E3 and B2.5 mandates, and Thailand, which has E10 and B3 mandates.[35]

Many observers continue to look to cellulosic biofuels as an area for future growth, but the U.S. Environmental Protection Agency (EPA) reduced the U.S. production target of cellulosic biofuels for the second straight year.[36] Cellulosic ethanol is made from biomass that is otherwise considered waste, such as woodchips, pulp, husks, and stems, or from low-value crops like switchgrass and jatropha. The final target for 2011 will be 25 million liters rather than the 950 million liters originally required by the Renewable Fuel Standard under the 2007 Energy Independence and Security Act.[37] Cellulosic ethanol can be harvested from degraded

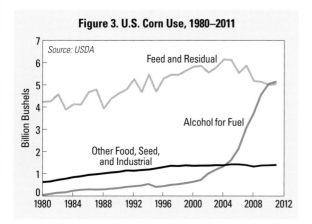

Figure 3. U.S. Corn Use, 1980–2011

University of Missouri's Food and Agricultural Policy Research Institute estimates that eliminating the 54¢ import tariff and 45¢-per-gallon blenders' credit would reduce industry profits by 7 percent and margins by 20 percent.[42] The Renewable Fuel Standard provides a guaranteed market of 50 billion liters in 2012, but the industry nearly matched this in 2010, suggesting that the mandate alone would not support the existing market.[43] The corn ethanol mandate increases to 57 billion liters by 2015.[44]

Brazilian sugarcane ethanol will likely become more prevalent in the United States if American ethanol subsidies and tariffs are removed, but poor weather and aging cane plants indicate that Brazilian sugar yields could fall in 2011.[45] Sugar ethanol is cheaper and more efficient to produce, although there are worries that its production may indirectly lead to deforestation.[46] Brazil's plans to build 103 new mills by 2019 (increasing capacity by 66 percent), the growing number of blending mandates around the world, and a global economic rebound all presage increased Brazilian ethanol production and exports in the coming years.[47]

land unsuitable for food, and it is often associated with lower greenhouse gas emissions because the crops grown for it are typically stronger greenhouse gas sinks and can convert more efficiently from feedstock to ethanol.[38] The EPA's target reduction reflects the technical challenges and costs of commercializing second-generation biofuels like cellulosic ethanol.[39]

In 2011, ethanol is projected to overtake the animal feeding industry as the largest corn consumer in the United States, helping production margins but shining a brighter light on ethanol subsidies.[40] (See Figure 3). A Senate bill will come before Congress that would cut ethanol production subsidies but maintain tax credits for infrastructure like refilling stations.[41] The

Sam Shrank was a MAP Sustainable Energy Fellow and Farhad Farahmand was a Climate and Energy research intern at Worldwatch Institute. The authors would like to thank Walter Falcon for his help in preparing this article.

Global Hydropower Installed Capacity and Use Increase

Matt Lucky

Global use of hydropower increased by over 5.3 percent between 2009 and 2010 (see Figure 1), reaching 3,427 terawatt-hours (TWh) by the end of that year.[1] The world's total consumption of hydropower increased each year between 2003 and 2010.[2] It also increased by at least 3.5 percent annually during five of the seven years between 2003 and 2010.[3]

Hydropower accounted for 16.1 percent of electricity use and 3.4 percent of energy use worldwide in 2010.[4] Only 12.5 percent of electricity came from hydroelectric power plants located in countries that belong to the Organisation for Economic Co-operation and Development.[5] All other countries were more dependent on hydropower in 2010, as it accounted for 19.8 percent of their electricity consumption.[6]

Global installed capacity of hydropower also experienced increases over the past several years. Installed capacities increased by 30 gigawatts (GW), or about 3 percent a year, in 2008, 2009, and 2010.[7] At the end of 2010, the total global installed capacity of hydropower was 1,010 GW.[8]

Although hydropower is produced in about 150 countries, it is concentrated in just a few countries and regions.[9] The Asia-Pacific region alone accounted for 31.8 percent of all hydropower production in 2010.[10] (See Figure 2.) Europe and Eurasia, South and Central America, and North America are the next highest producing regions, at 25.3, 20.3, and 19.3 percent, respectively.[11] Africa, at 3.0 percent, and the Middle East, at 0.4 percent, account for the least production worldwide.[12] Africa is considered the region with the most potential for increased hydropower production.

Just five countries accounted for approximately 52 percent of the installed hydropower capacity worldwide at the end of 2010.[13] China

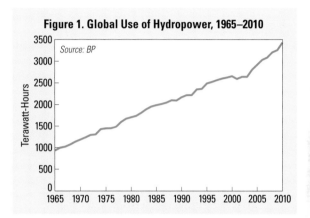

Figure 1. Global Use of Hydropower, 1965–2010

Source: BP

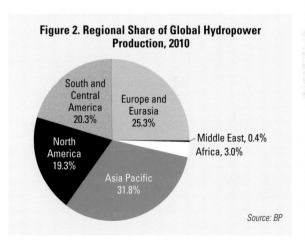

Figure 2. Regional Share of Global Hydropower Production, 2010

South and Central America 20.3%

Europe and Eurasia 25.3%

Middle East, 0.4%

Africa, 3.0%

North America 19.3%

Asia Pacific 31.8%

Source: BP

had the most, at 213 GW.[14] Brazil was next, at 80.7 GW, followed by the United States (78 GW), Canada (75.6 GW), and Russia (55 GW).[15] The 27 members of the European Union also had 130 GW of installed hydropower by the end of 2010.[16]

China also added more hydropower capacity than any other country in 2010, connecting 16

GW to the grid; it plans to add 140 more GW of capacity by 2015.[17] Brazil added 5 GW of capacity to its grid in 2010, and an additional 8.9 GW of hydropower was under construction.[18] Canada added half a gigawatt of hydropower in 2010, and an additional 11 GW was also under construction.[19] A total of $40–45 billion was invested in large hydropower projects worldwide in 2010.[20]

There are now three hydropower plants larger than 10 GW. The largest is at the Three Gorges Dam in China, which is rated at 18.2 GW; it is expected this power plant will be upgraded to 22.5 GW by 2012.[21] The next largest ones are the Itaipu Hydroelectricity Power Plant in Brazil, rated at 14 GW, and the Guri Dam in Venezuela, rated at 10.2 GW.[22] Brazil is also home to the world's fourth largest hydropower plant, the Tucuruí Dam, which is rated at 8.5 GW.[23]

Twelve countries generated more than 50 TWh of hydropower in 2010.[24] China generated by far the most, producing 721 TWh; this accounted for 17.1 percent of the country's total electricity use in 2010.[25] Brazil, Canada, and the United States were the next largest producers of hydropower in 2010. (See Figure 3.)[26]

Fifteen countries generated at least 90 percent of their electricity from hydroelectric

power plants in 2008.[27] (See Table 1.) Four of them—Albania, Bhutan, Lesotho, and Paraguay—generated all their electricity from hydro during that year.[28] Norway, which generated 94.8 percent of its electricity from hydroelectric power plants in 2010, consumed 117.9 TWh of hydropower in 2010, making it the sixth largest consumer of hydropower in the world.[29] Iceland, New Zealand, and Norway produced the most hydropower per person.[30]

A technology that has grown in importance over the past several years is small hydropower, defined as a plant with an installed generating capacity of less than 10 megawatts (MW). As of 2009, there were about 60 GW of installed small hydropower worldwide, so this accounted for less than 6 percent of all installed hydropower.[31] Fifty-five percent (33 GW) of this capacity was located in China. Japan (4 GW), the United States (3 GW), Germany (2 GW), India (2 GW), and Spain (2 GW) were the countries with the next greatest amount of installed small hydropower as of 2009.[32] Some $2 billion was invested in these projects worldwide in 2010, down 43 percent from 2009.[33]

Micro-hydropower, which is defined as a plant with an installed capacity of 100 kilowatt-hour (kWh) or less, is growing in popularity. It is effective for electrifying small communities that lie well off the grid, and it is also displacing small diesel generators in rural communities.

There are 136 GW of pumped storage hydropower operating worldwide.[34] Pumped storage hydropower is a process that uses excess or cheap electricity to pump water uphill into a reservoir, which is used later to generate electricity when demand is high and electricity is expensive. This form of energy storage is becoming increasingly popular in countries with high renewable penetration, as it complements variable generation technologies like solar and wind.[35] In 2010, pumped storage hydropower grew by 4 GW, and an additional 5 GW was also under construction.[36] Europe, Japan, and the United States (with 20.5 GW) have most of the world's pumped storage hydropower capacity.[37]

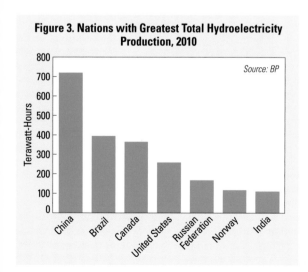

Figure 3. Nations with Greatest Total Hydroelectricity Production, 2010

Source: BP

Table 1. Countries with At Least 90 Percent of Total Electricity Production from Hydropower in 2008

Country	Percent
Albania	100.0
Bhutan	100.0
Lesotho	100.0
Paraguay	100.0
Mozambique	99.9
Nepal	99.7
Zambia	99.7
Dem. Republic of Congo	99.4
Burundi	99.0
Tajikistan	98.2
Angola	96.5
Belize	94.9
Norway	94.8*
Laos	92.5
Kyrgyzstan	90.9

*Norway data from BP.
Source: DOE, EIA, International Energy Statistics (Washington, DC: 2011).

Hydropower provides many ancillary benefits. The damming of rivers provides flood control, recreational opportunities, and a source of irrigation and water supply. Hydropower plants can also be ramped up and down very quickly, making this a flexible source of electricity that can be used to meet both baseload and peakload energy demand. Generating hydropower also has no direct greenhouse gas (GHG) emissions.

There are several negatives associated with hydropower, however. Damming interrupts the flow of rivers, altering water quality both upstream and downstream. Sediment buildup in reservoirs harms wildlife and often shortens the life of hydro plants. Fish passages are also disturbed, leading to higher fish mortality rates. There are significant GHG emissions associated with the flooding of valleys to create reservoirs. In addition, the creation of the world's largest dams has displaced considerable numbers of people, which makes small hydropower even more appealing.

Although hydropower largely complements renewable energies, it has recently come into direct competition with them in isolated cases. High levels of hydropower production led to curtailments of wind production in the northwestern United States in May 2011.[38] In the United States, federal regulations limit how much water can be spilled over a dam, as this can disrupt water quality and harm endangered fish.[39] As a result, utilities often have no choice but to operate hydropower over wind power during periods of low electricity demand.[40] Unfortunately, this has reduced production tax credits received by wind generators, as they are not producing electricity even when the wind is blowing.[41]

One quality that makes hydropower a very competitive source of electricity is its price. The average cost of electricity from a hydropower plant larger than 10 MW is 3–5¢ per kWh.[42] Electricity from a small hydropower plant costs 5–12¢ per kWh.[43] Micro-hydropower generates electricity at a cost of 7–40¢ per kWh.[44]

Hydropower will likely continue to grow in the future. Its competitive price and climate benefits will make it an attractive option as countries attempt to lower their GHG emissions. Very large hydropower projects in China and Brazil will likely account for a majority of this growth, but small hydropower is also likely to expand, especially as populous countries like India continue to pursue rural electrification.

Matt Lucky is a MAP Sustainable Energy Fellow at Worldwatch Institute.

Energy Poverty Remains a Global Challenge for the Future

Michael Renner and Matt Lucky

Acknowledging that many people in developing countries do not have access to affordable, reliable, and safe sources of energy, the United Nations General Assembly designated 2012 to be the International Year for Sustainable Energy for All.[1] Expanding access to modern energy services for lighting, heating, refrigeration, cooking, water pumping, communications, and other services is essential for reducing poverty, improving health and education, increasing incomes, and enhancing rural livelihoods.[2] It will be difficult to achieve a number of the internationally endorsed Millennium Development Goals without improving energy access.[3]

With greater energy access, long hours spent collecting fuelwood can instead be directed toward education or income-generating activities. With more reliable lighting, stores and other businesses can stay open into the evening, allowing more economic transactions. Children are able to study after sunset. People's health improves when the deadly air pollution associated with conventional fuels is eliminated and when health clinics can count on reliable sources of electricity.

Improving energy access is important, but measuring it and facilitating the proper policy measures to increase access are difficult. Energy access means many things to many people—access to cooking fuels, transportation, electricity, and heating and cooling, to name just a few—and not all organizations and governments agree on how to measure it. Single indicator statistics, such as a population's access to electricity, are simple to collect and can be very powerful, as they give comparable and easily interpreted statistics across regions. Multidimensional indicators, which measure more than one set of data, can be more complex and difficult to compare across regions as they involve several indicators.[4]

The advantage of multidimensional indicators, however, is that they are more representative of an entire region's energy access situation and they can drive more all-encompassing policy change.

While many organizations publish statistics on energy access, it is important to understand their definitions and limitations. According to the International Energy Agency (IEA), modern energy access is defined as "a household having reliable and affordable access to clean cooking facilities, a first connection to electricity and then an increasing level of electricity consumption over time to reach the regional average."[5] The United Nations Advisory Group on Energy and Climate Change defines it as "access to clean, reliable and affordable energy services for cooking and heating, lighting, communications and productive uses."[6]

The IEA measures electrification rates by looking at who has access to electricity in their household. This includes people who buy electricity both on-grid and off-grid, and it also includes those who self-generate electricity and have been surveyed by their governments. Not included are people who are illegally connected to the grid and do not pay for their electricity.[7]

As the United Nations promotes the International Year for Sustainable Energy for All, important definitional questions linger. While having a legal connection to the grid may technically qualify a household as having access to electricity, if that home has only a couple of lightbulbs and unreliable access to the grid, policymakers must ask themselves if that is really what they are striving for. Should readily available access to transportation services factor in? Does supply meet demand? Are energy services reliable? Is energy affordable for all? These are all important questions that need to be considered when discussing energy access.

According to estimates by the IEA, more than 1.3 billion people are currently without any access to electricity.[8] The United Nations estimates that another 1 billion people have unreliable access.[9] At least 2.7 billion people, and possibly more than 3 billion, lack access to modern fuels for cooking and heating.[10] (See Figure 1.)

The largest populations currently lacking access to electricity live in sub-Saharan Africa and South Asia. The two regions combined account for more than 80 percent of the people worldwide without electricity.[11] (See Table 1.) The problem is most pronounced in rural areas of sub-Saharan Africa, where the electrification rate is just 14 percent.[12] This has important socioeconomic ramifications; these low electrification rates have encouraged urban migration, as higher electrification rates in cities are associated with more economic opportunity.[13] Half of South Asia's rural population, in contrast, has access to electricity.

In absolute terms, India is the country with by far the most people—289 million—without access to electricity.[14] Another six countries have at least 50 million people who lack access, and seven more have at least 20 million people in that situation.[15] (See Figure 2.) Unlike the rest of the world, in Africa grid expansion is not keeping pace with population growth. Unless policies change, the continent's population without access to electricity is projected to grow to 630 million people in 2015 and to 700 million by 2030.[16] The current share of the population with access to electricity in Africa ranges from a high of 99 percent in Mauritius and 75 percent in South Africa to a low of 9 percent in Uganda and Malawi.[17]

In Asia, Singapore has achieved universal access, and Brunei, China, Malaysia, Thailand, and Taiwan have electrification rates of 99 percent or more.[18] Vietnam is close to 98 percent, and the Philippines, close to 90 percent.[19] But at the other end of the spectrum, Myanmar's rate is just 13 percent, and Cambodia's is 24 percent.[20] Among South Asian nations, Afghanistan has by far the lowest rate, at less than 16 percent.[21] Electrification rates in Latin America are gener-

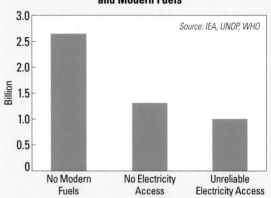

Figure 1. Number of People Lacking Access to Electricity and Modern Fuels

Source: IEA, UNDP, WHO

Table 1. Lack of Electricity Access, by Region, 2009

Region	Population without Electricity			
	Total	Total	Urban	Rural
	(million)	(percent)	(percent)	(percent)
Africa	587	58	31	75
North Africa	2	1	0	2
Sub-Saharan Africa	585	70	40	86
Developing Asia	675	19	6	27
China and East Asia	182	9	4	14
South Asia	493	32	11	40
Latin America	31	7	1	26
Middle East	21	10	1	28
All Developing Countries	1,314	25	9	37

Source: IEA, "Access to Electricity," World Energy Outlook website.

ally quite high (93.2 percent overall), but Bolivia, Honduras, and Nicaragua all fall below 80 percent.[22] Haiti (at 39 percent) has by far the lowest rate in the region.[23] Countries in the Middle East have virtually universal access, although Iraq's rate (86 percent) is less impressive and Yemen (40 percent) is a big exception.[24]

Between 1970 and 1990, more than 1 billion people—half of them in China—gained access to electricity services. Then from 1990 to 2008,

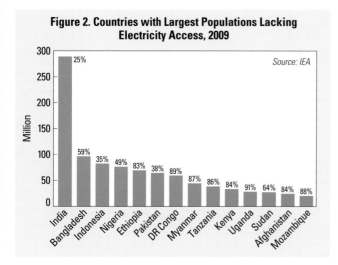

Figure 2. Countries with Largest Populations Lacking Electricity Access, 2009

Source: IEA

cer.[29] An estimated 44 percent of those who die are children; among adult deaths, 60 percent are women.[30]

The largest populations relying on traditional biomass for cooking are found in the developing regions of Asia, with by far the largest number (836 million in 2009) in India and more than 100 million each in Pakistan, Bangladesh, and Indonesia.[31] Altogether, 54 percent of the total population of developing Asia relies on traditional biomass.[32] In Africa, the absolute number is 657 million people, but the share is higher: 65 percent.[33] In Ethiopia, the Democratic Republic of Congo, and Tanzania, more than 90 percent of the population uses traditional biomass.[34] In Asia, only Myanmar has a similarly high share.[35] (See Table 2.)

According to the IEA, an estimated $9.1 billion was invested worldwide in 2009 in extending access to modern energy services.[36] This helped provide an additional 20 million people with electricity access and 7 million people with advanced biomass cookstoves.[37] The IEA projects that from 2010 to 2030, an annual average of $14 billion will be spent, mostly on urban grid connections (and a considerable amount will still go to fossil fuel power plants and large-scale hydropower).[38]

The projected funding will likely still leave 1 billion people—largely those who live in the most remote parts of developing countries—without electricity and 2.7 billion without clean cooking facilities in 2030.[39] The IEA estimates that average annual investments will need to rise to $48 billion to provide universal modern energy access.[40] For 2010–30, cumulative spending needs to reach an estimated $960 billion—which is $640 billion more than current annual investments would provide.[41]

In addition, the cumulative investment needed to provide more than 250 million households worldwide with advanced biomass cookstoves is estimated at $17 billion, while $37 billion is required for biogas systems for 70 million households and $20 billion for liq-

close to 2 billion more people secured access.[25] China, Vietnam, Thailand, Sri Lanka, South Africa, and Brazil are among the countries that have had considerable success in expanding rural access since the 1990s, principally through large-scale grid extension projects that have relied on fossil-fuel-based generating technologies or large-scale dams.[26] Brazil's Luz para Todos (Light for All) program, for instance, was initially mostly focused on hydropower-driven grid extension; decentralized renewable energy came into play only later on, in remote areas.[27]

Hundreds of millions of people continue to rely on traditional biomass—firewood, charcoal, manure, and crop residues—for many energy services. As the U.N. Environment Programme notes, "apart from being energetically inefficient … the use of traditional bioenergy is connected to several severe health and environmental problems," including indoor air pollution, forest and woodland degradation, soil erosion, and black carbon emissions that contribute to global warming.[28]

Indoor air pollution from cooking and heating with traditional sources of energy is exacting a massive toll on human health. Worldwide, it causes almost 2 million deaths annually—nearly all of them in the developing world—from pneumonia, chronic lung disease, and lung can-

Table 2. Populations Relying on Traditional Use of Biomass for Cooking, 2009

	Population	
	(million)	(percent)
Africa	657	65
Nigeria	104	67
Ethiopia	77	93
DR Congo	62	94
Tanzania	41	94
Kenya	33	83
Developing Asia	1,921	54
India	836	72
Bangladesh	143	88
Indonesia	124	54
Pakistan	122	72
Myanmar	48	95
Latin America	85	19
Middle East	0	0
Developing Countries	2,662	51

Source: IEA, Energy for All—Financing Access for the Poor *(Paris: 2011).*

uefied petroleum gas stoves for nearly 240 million households.[42]

A growing number of governments, international agencies, nongovernmental organizations, and businesses are working to overcome energy poverty, particularly by focusing on renewable energy. To date, 68 developing-country governments have adopted formal targets for improving access to electricity.[43] A far smaller number have targets for providing access to modern fuels (17) and improved cook stoves (ICS) (11).[44]

Three general options exist for increasing rural access to electricity: connecting communities to a centralized grid, installing decentralized micro-grids, and constructing off-grid power systems for individual homes and buildings.[45] Access to electricity in rural areas is unlikely to be provided through the first option, as many countries lack the necessary infrastructure and funding. More feasible are mini-grids that link a community to a small central generating plant located nearby and run by a village cooperative or an individual entrepreneur. Decentralized off-grid options are often the only realistic option for the most remote rural locations, where people are not numerous enough for mini-grids or are too poor to construct them.

Small hydropower plants, small wind power installations, village or household-size biogas plants, solar home systems, and pico solar (solar lanterns and lights) can be cost-effective options for mini-grid or off-grid energy solutions. Hybrid systems mix a range of options, often tied into a village-scale mini-grid of anywhere from 10 to 1,000 kilowatts.[46]

Improved cook stoves play an important role, since they allow people either to make use of more modern fuels or to use traditional fuels much more efficiently. ICS can double or triple the thermal efficiency of traditional fuels, thereby reducing dangerous indoor air pollutants that cause disease and premature deaths and reducing black carbon emissions that contribute to climate change.[47] Reducing or eliminating the long hours that women and children spend foraging for fuelwood frees up valuable time and reduces pressures on forests and ecosystems.[48] Consuming less fuel also saves money, giving people more disposable income and allowing them to invest more in their futures.

Only a fraction of the people who need them—roughly 830 million—are estimated to have access to ICS, most of whom are in China.[49] In sub-Saharan Africa, only 34 million people have access to such stoves, according to a joint report by the World Health Organization and the U.N. Development Programme.[50] But these numbers are only estimates, as half of all developing countries do not have reliable ICS data.

The International Year for Sustainable Energy for All highlights the grave reality of energy access in today's world. Unfortunately, billions of people still live without access to electricity and modern cooking fuels and technologies. If funding is not increased substantially, this situation will not improve in the future.

Michael Renner is a senior researcher and Matt Lucky is a MAP Sustainable Energy Fellow at Worldwatch Institute.

Transportation Trends

Smiley.toerist

A work train delivers ballast on the high-speed rail line to Malaga, Spain

For additional transportation trends, go to vitalsigns.worldwatch.org.

Auto Industry Stages Comeback from Near-Death Experience

Michael Renner

Following a precipitous plunge in 2008 and 2009, the world's auto industry saw production and sales soar in 2010. According to London-based IHS Automotive, passenger-car production rose from 45.3 million in 2009 to 54.9 million in 2010.[1] (See Figure 1.) Including so-called light trucks (which are also used to transport passengers), the numbers shot up from 60 million to a new record of 74.7 million.[2] Global sales of all light vehicles increased from 72.2 million in 2009 to 75.4 million in 2010.[3]

About 50 countries worldwide produce cars.[4] Seventeen of them each made at least 1 million light vehicles in 2010.[5] South Korea (4.2 million) and India and Brazil (3.2 million each) continue to rise up the ranks, whereas once-strong second-tier producers like the United Kingdom, France, Italy, and Canada are losing ground.[6]

The top four producers—China, Japan, the United States, and Germany—are in a league of their own, together accounting for 53 percent of global output.[7] The pecking order among them, however, has changed dramatically in just a handful of years. (See Figure 2.) Japan overtook the United States as the largest producer in 2006, but it held that distinction for just three years, at which point China's production shot up.[8] Although still behind in terms of technological sophistication, China is accelerating the volume of cars it churns out. Its light-vehicle production has more than tripled since 2005, reaching 16.8 million in 2010—as many vehicles as produced in Japan and the United States combined.[9]

Japan's auto industry took a powerful hit when the earthquake and tsunami disaster hit in March 2011. By the end of that month, forced plant shutdowns led to a cumulative reduction in output of about 400,000 vehicles.[10] Hundreds of car-parts suppliers are located near the disaster zone. The resulting shortages of auto components may continue to reverberate in Japan and even abroad for months, since that country is the second-largest parts exporter after Germany and accounts for 11 percent of global components production.[11] Japan's largest producer, Toyota, does not expect a return to normal production levels before the end of 2011.[12]

IHS Automotive projects that Japan's output will fall from 9.3 million in 2010 to 6.6 million in 2011.[13] This would trim the country's share of global production from 12.5 percent in 2010 to 8.5 percent of a projected 2011 total of 76.4 million.[14]

Worldwide, industry analysts expect assembly of light vehicles to stage strong growth in coming years. PricewaterhouseCoopers (PWC) Autofacts projects output of 93.5 million by 2015.[15] They expect China's output to reach 21.4 million units, far ahead of the United States at 10.5 million and Japan at 9.2 million.[16] India's output is projected to rival that of Germany at about 6 million.[17] Other nations whose production is expected to expand strongly are Brazil, Thailand, Russia, and Argentina.[18]

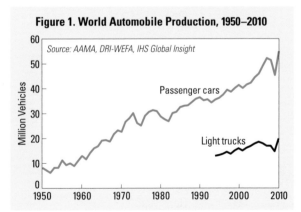

Figure 1. World Automobile Production, 1950–2010

Source: AAMA, DRI-WEFA, IHS Global Insight

Analysts expect sales to reach a new peak in 2011. IHS Automotive projects a global figure of 82.8 million.[19] Market saturation and economic crisis are limiting European, Japanese, and North American sales. The pace is being set by the so-called BRIC nations—Brazil, Russia, India, and China. Together, BRIC sales are expected to surpass the combined sales in Western Europe and Japan in 2011.[20] Except in Russia, new sales records were set in these countries in 2010—9.52 million units in China, 2.65 million in Brazil, and 1.84 million in India.[21]

Car sales in China soared 50 percent in 2009 and 33 percent in 2010, but growth rates in 2011 are expected to be lower.[22] This is due to the government's decision to raise sales taxes for small cars back to pre-economic crisis rates, along with a move to restrict the number of new license plates issued in Beijing in 2011 in a bid to reduce the traffic that threatens to suffocate that city.[23] Beijing had 4.7 million cars at the end of 2010, with more than 700,000 added in that year alone.[24] By 2017, Chinese car sales are conservatively forecast to exceed 23 million vehicles—roughly as many cars as will be sold then in the United States, Germany, and Japan combined.[25]

India also saw car sales grow 33 percent in 2010. The launch of low-cost vehicles like the Tata Nano leads analysts to believe that the Indian market will for several years grow at even higher rates than the market in China.[26] In 2010, at 3.5 million units sold, Brazil surpassed Germany to become the world's fourth largest car market.[27]

An estimated 669 million passenger cars are on the world's roads.[28] When light- and heavy-duty trucks are included, the number rises to 949 million vehicles.[29] Though their fleets are expanding rapidly, Brazil, China, and India still have far fewer cars per person than western countries do: 143 vehicles per 1,000 people in Brazil, 40 per 1,000 in China, and 14 per 1,000 in India—compared with 673 per 1,000 in the Group of Seven industrial countries.[30]

The global auto industry continues to be dominated by a relatively small number of com-

Figure 2. Light Vehicle Production, Leading Countries, 1995–2010

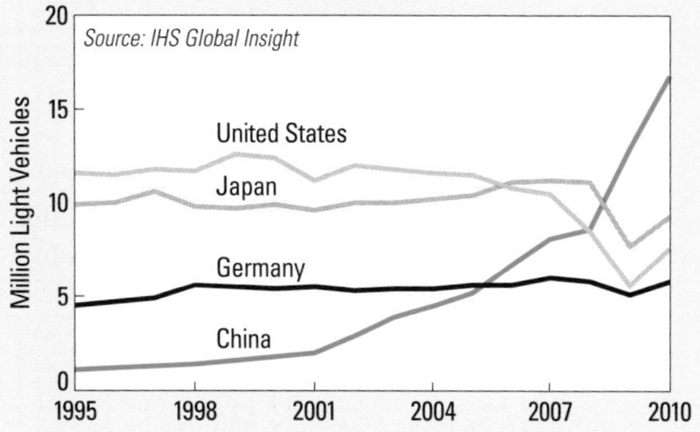

Source: IHS Global Insight

panies, principally from North America, Europe, and Japan. Different sources report conflicting corporate production numbers, in part because they classify various types of vehicles into different categories. Also, some analysts focus strictly on individual companies, while others look at broader "alliance groups" that include subsidiaries and affiliated firms.[31]

If the broader corporate alliances are considered, VW-Suzuki had the upper hand in 2010, with just under 10 million units compared with Toyota's 9.3 million.[32] The only company in the top 10 not from the traditional car-manufacturing countries is South Korea's Hyundai Group.[33] The top 20 includes India's Tata Group, Iran's SAIPA, and the Chinese companies Chang'an, Geely, Chery, BYD, and FAW.[34]

Foreign car companies that entered China from the 1990s onward were required to achieve a high level of domestic content—typically, 70 percent within three years of entry.[35] But bona fide Chinese manufacturers still have a long way to go to catch up, even at home. International joint ventures in China accounted for 84 percent of output at the end of the last decade.[36] And European, Japanese, and U.S. multinationals also retain a strong sales presence in the Chinese market, although Chinese firms account for 42 percent of sales.[37]

Since the mid-1980s, the auto industry has become increasingly globalized. But an even more pronounced trend has been toward regional integration, with car companies' respective home regions continuing to play an important role for production and sales. Within individual regions, there has been a shift toward lower-cost manufacturing locations, such as Mexico and southern states in the United States, Spain, Eastern Europe, Southeast Asia, and China.[38]

The International Council on Clean Transportation notes that national policies regulate about 70 percent of the world's transport-related emissions through efficiency standards.[39] The United States, Canada, the European Union (EU), Japan, China, and South Korea now all have mandatory standards.[40] Australia's remain voluntary.[41] Mexico, India, Indonesia, and Thailand are in the process of drawing up regulations.[42]

Fuel efficiency has been improving in all the major car nations since 2002, and enacted, proposed, and studied limits will bring about additional reductions in the amount of fuel needed to travel a given distance.[43] (See Figure 3.) At the same time, substantial differences among individual countries remain. Japan and the EU continue to be the leaders in fuel efficiency.[44] The United States, Canada, and Australia—nations that have traditionally been heavily car-dependent and have enjoyed low gasoline prices—are still lagging behind.[45]

Rising gasoline prices have driven some improvement in what has long been a dismal fuel economy performance in the United States.[46] For the 2004 model year, just 1.1 percent of all light vehicles sold achieved between 30 and 35 miles per gallon (mpg), and just 0.5 percent were above 35 mpg.[47] That improved to 12 percent and 3 percent, respectively, for the 2010 model year, largely because the share of sports utility vehicles and other light trucks declined by 12 percent during this period.[48]

In 2009, light vehicles purchased in the United States averaged emissions of 247 grams of carbon dioxide (CO_2) per kilometer traveled, down from 423 grams in 1975.[49] Although this was a substantial improvement, the United States still lags far behind Japan (131 grams) and the 27-member European Union (average of 146 grams, with 134 grams in France, 150 grams in the United Kingdom, and 154 grams in Germany).[50] The EU has a 2015 target of 130 grams.[51]

Among manufacturers, Italian and French companies, along with the Japanese companies Toyota and Honda and the Korean firms Hyundai and Kia, fare best.[52] The U.S. companies Chrysler, Ford, and GM are among the worst performers.[53] Interestingly, in Europe, where policy is more strongly in favor of efficient, low-carbon vehicles, all companies offer better-performing models than in the United States, where such policies remain comparatively weak. The discrepancy is most pronounced for Ford, GM, Daimler, and BMW.[54]

There is continued interest in developing alternative propulsion vehicles. Half of the world's purchases of so-called hybrid-electric cars—close to 1 million in 2010—were made in Japan, where such vehicles accounted for 11 percent of all sales.[55] (See Table 1.) The United States was once the leading market, but hybrid sales there peaked in 2007, and the country fell behind Japan in both absolute numbers and percent of total sales.[56] In Europe, where fuel-efficient diesel vehicles have strong appeal, the numbers and market share of hybrids are much smaller.[57]

Figure 3. Automobile Fuel Consumption, 2002–10, with Projections to 2020

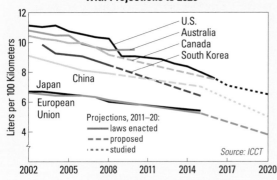

Table 1. Worldwide Hybrid Vehicle Sales, 2010

	Hybrid Vehicle Sales	Hybrid Market Share
		(percent)
Japan	492,000	11.0
United States	291,000	2.5
Europe	110,000	0.7
Rest of the World	90,400	0.7
World	983,400	2.2

Note: The table includes sales of electric vehicles, but they are minute compared with hybrids.
Source: "Hybrid Car Statistics," undated, at www.all-electric-vehicles.com/hybrid-car-statistics.html.

Toyota introduced hybrid vehicles in 1997, and by February 2011 the company had sold a cumulative 3 million units, principally of the flagship Prius model.[58] The company remains the dominant producer of hybrids worldwide. It sold 690,000 of these cars in 2010.[59] It had aimed to reach 1 million in 2011, but the earthquake and tsunami tragedy may prevent Toyota from reaching this goal.[60]

China's Energy Saving and New Energy Vehicle Development Plan for 2011–20 is aimed at boosting plug-in hybrids and electric vehicles by investing some 100 billion yuan over the next decade.[61] By 2015, the goal is to have 1 million such vehicles on the road.[62]

Forecasts on hybrid and electric vehicles continue to be mostly ambitious. PWC Autofacts projects annual production to climb to about 9 million units by 2020, which would be 9 percent of global car production then.[63] The firm expects that a third of the total will be plug-in and battery-electric vehicles.[64] Even more optimistically, Carlos Ghosn, the CEO of Nissan-Renault, predicted in 2010 that electric cars alone will constitute 10 percent of global sales by 2020.[65] Similarly, a widely cited report titled *Sizing the Climate Economy* by HSBC Global Research projected, as part of an optimistic scenario, that global sales of electric vehicles would rise to 8.7 million by 2020 and those of plug-in hybrid electric vehicles would reach 9.2 million.[66]

Other observers are more skeptical. A J.D. Power and Associates study in October 2010 said that combined hybrid and electric vehicle sales of 5.2 million units might capture just slightly more than 7 percent of global car sales in 2020.[67] A 2011 report by the firm that focused on the U.S. market concluded that hybrid and electric sales would remain below 10 percent through 2016.[68]

A review of scenarios and assessments by various analysts concludes that gas-electric hybrids will penetrate global markets sooner and more easily than full-fledged electric vehicles (plug-in hybrids and electric vehicles).[69] Most scenarios discount the likelihood that electric vehicles will surpass 25 percent of new sales even by 2050 in the absence of sufficient technological improvements, cost reductions, and strong policy incentives or regulations.[70] And of course electric vehicles offer climate benefits only if the power sector moves away from carbon-intensive fuels.

Michael Renner is a senior researcher at Worldwatch Institute.

High-Speed Rail Networks Expand

Michael Renner

Interest in high-speed rail (HSR) is growing around the world, with the number of countries running such trains expected to grow from 14 in mid-2011 to 24 over the next few years.[1] Although there is no single definition of high-speed rail, the threshold is typically set at 250 kilometers per hour (km/h) on new dedicated tracks and 200 km/h on existing, upgraded tracks.[2]

The length of high-speed rail tracks worldwide is undergoing explosive growth. In 2009, some 10,700 kilometers of track were operational.[3] Just two years later, the total length has grown to almost 17,000 kilometers.[4] Another 8,000 kilometers of track are currently under construction, and about 17,700 kilometers more are planned, for a combined total of close to 43,000 kilometers.[5] That is equivalent to about 4 percent of all rail lines—passenger and freight—in the world.[6]

By track length, the current high-speed leaders are China, Japan, Spain, France, and Ger-

many.[7] (See Figure 1.) Planned new routes in France and Spain will put those two countries ahead of Japan in coming years.[8] (See Table 1.)

Other countries are joining the high-speed league. Turkey has ambitious plans to reach 2,424 kilometers and surpass the length of Germany's network.[9] Italy, Portugal, and the United States all hope to reach track lengths of more than 1,000 kilometers.[10] Another 15 countries have plans for shorter networks.[11]

The number of trains running on these tracks is expanding as well. In January 2008, a total of 1,737 high-speed train sets were in operation worldwide.[12] By January 2011, the global fleet had grown to 2,517.[13] Two thirds of the fleet is found in just five countries: France (494 train sets), followed by China (406), Japan (359), Germany (250), and Spain (203).[14] Another 800 train sets will be added by these five countries in the next two years.[15] Altogether, the global fleet is expected to total more than 3,700 units by 2015.[16]

Track length and number of trains alone do not necessarily indicate how heavily high-speed rail is actually being used. Worldwide, travel volume increased from about 30 billion passenger-kilometers (pkm) in 1970 to more than 130 billion pkm in 2000 and at least 200 billion pkm today.[17] The International Union of Railways (UIC) offers a breakdown for leading countries (although it does not include data for China in its global total).[18] (See Table 2.) Worldwide, the high-speed segment accounts for a still small 7 percent of all rail passenger-kilometers traveled in 2010.[19]

Interest in high-speed rail goes back more than a century. In 1903, an experimental German train reached a speed of 203 km/h, but it would be another 60 years before regular high-speed service was introduced in Japan in 1964.[20]

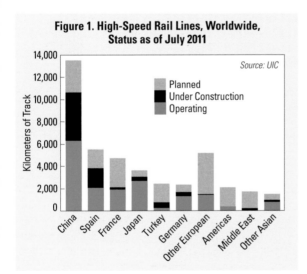

Figure 1. High-Speed Rail Lines, Worldwide, Status as of July 2011

Source: UIC

Planned
Under Construction
Operating

Kilometers of Track

China, Spain, France, Japan, Turkey, Germany, Other European, Americas, Middle East, Other Asian

Within a dozen years of the introduction of Japan's "Shinkansen" trains (known as bullet trains in the West), the milestone of 1 billion passengers served was reached.[21] Today, these trains carry more than 350 million passengers every year—more people than all the other high-speed lines in the world combined.[22] The high volume is the result not just of high-capacity trains (which include double-deckers) but also extensive geographic coverage, very frequent and regular departures, and an impeccable on-time arrival performance.[23]

Japan's success helped revive interest in high-speed rail in Europe. France was the second country to build an HSR system, with an inaugural line from Paris to Dijon opening in 1981.[24] Germany and Italy also started to build high-speed lines. HSR travel in member states of the European Union grew from 15.2 billion passenger-kilometers in 1990 to 97.6 billion pkm in 2008, equaling 24 percent of total intercity rail travel in that region.[25] For 2009, UIC reports that the figure grew to 104.4 billion pkm, even during the world economic crisis.[26]

France continues to account for about half of all European high-speed rail travel.[27] High-speed reached an astounding 62 percent of the country's entire passenger rail travel volume in 2008, up from just 23 percent in 1990.[28] This is due to an impressive network and affordable ticket prices. As rail expert and blogger Yonah Freemark explains: "In France, SNCF [the national rail operating company] was conceived as a service providing more than simply transportation, but also societal equalization, befitting its designation as a public enterprise…. The operator offers high-speed tickets at cheap prices compared to comparable services in Germany, Spain, or Italy—a policy that has resulted in highly frequented TGV trains and a general equality in train service for people of all income classes."[29]

A draft plan for French transportation infrastructure investments for the next two decades allocates 52 percent of $236 billion to high-speed rail.[30] (Another 32 percent will go to urban trams, subways, and bus lines, in a strong overall backing for public modes of transporta-

Table 1. High-Speed Rail in Operation, Under Construction, and Planned, as of July 2011

	In Operation	Under construction	Planned	Total
	(kilometers of track)			
China	6,299	4,339	2,901	13,539
Spain	2,056	1,767	1,702	5,525
France	1,896	210	2,616	4,722
Japan	2,664	378	583	3,625
Turkey	235	510	1,679	2,424
Germany	1,285	378	670	2,333
Other European	1,400	72	3,717	5,189
Americas	362	XX	1,726	2,088
Middle East	XX	200	1,505	1,705
Other Asian	757	186	544	1,487
World	16,954	8,040	17,643	42,637

Source: UIC, Km of High Speed Lines in the World.

Table 2. High-Speed Rail Passenger Travel, 2010

Country	Distance
	(billion passenger-kilometers)
Japan	76.0
France	51.9
Germany	23.9
Spain	11.7
Italy	11.3
South Korea	11.0
Taiwan	7.5
Sweden	2.8
Other Europe	4.9

Source: UIC, Km of High Speed Lines in the World.

tion; just 5 percent will go to roads and airports, and the remainder to ports and waterways.)[31]

High-speed rail lines in France helped connect previously isolated regions of the country and provided a spur to their economic development. A number of lines in other countries have been explicitly planned with this goal in mind, including the Madrid-Seville route in Spain as well as a proposed line linking Amsterdam and

Groningen in the Netherlands.[32] Likewise, a Morgan Stanley analysis of China's high-speed rail system notes that: "Many economically challenged cities in west and central China will be revitalized because of the hub effect created by the HSR system."[33]

Spain only entered the high-speed rail world in 1992, but its network grew rapidly to become the longest in Europe and will expand to be the second largest in the world, after China's.[34] In 2005, the Spanish government announced an ambitious plan for some 10,000 kilometers of high-speed track by 2020, which would allow 90 percent of Spaniards to live within 50 kilometers of an HSR station.[35]

Like Spain, China was a latecomer to the world of high-speed rail. But with a massive planning and investment effort, it climbed the ranks of the world's high-speed league. Currently China is investing about $100 billion annually in railway construction.[36] The share of railway infrastructure investment allocated to HSR has risen from less than 10 percent in 2005 to a stunning 60 percent in 2010.[37] By 2012, China plans to have built an impressive 42 different passenger lines and to have about 17,000 kilometers of high-speed track in operation.[38]

China's gargantuan effort has been tarnished by corruption and reports of poor quality of track construction, however, which in February 2011 led to the firing of Railway Minister Liu Zhijun.[39] Safety concerns subsequently led to a decision to reduce the top speeds of most trains from 350 km/h to 300 km/h.[40] In July 2011, some 40 people died and 191 were injured in a crash near Wenzhou.[41] Following the accident, China's government decided to suspend approvals of new high-speed lines and to re-evaluate safety systems on rail projects that have already been approved but not yet built.[42]

The Wenzhou crash was the second fatal HSR event, following a derailing in Germany in 1998 that killed 101 people and injured more than 100.[43] Still, high-speed rail travel is very safe. Japanese high-speed trains in particular have an unblemished safety record, with no passenger casualties.[44]

High-speed rail has become an attractive and reliable travel option at a time when air travel becomes ever more cumbersome and time-intensive due to security concerns. The number of passengers carried by modern high-speed train sets easily rivals that of even the largest airplanes. The capacity of high-speed trains currently in operation ranges from a low of 160 to a high of 1,324 seats.[45]

Japan's Shinkansen trains are known for their exceedingly high degree of reliability. JR Central, the largest of the Japanese rail operating companies, reports that the average delay per high-speed train throughout a year is an incredible half a minute.[46] The Shinkansen are formidable competitors on distances up to about 900 kilometers.[47] (See Table 3.) On all routes in Japan where both air and high-speed rail connections are available, rail has captured a 75 percent market share.[48]

Other countries have had similar experiences with high-speed rail networks. In France, rail's market share of the Paris-Marseille route rose from 22 percent in 2001, before the introduction of high-speed service, to 69 percent in 2006.[49] On the Madrid-Seville route, rail's share rose from 33 to 84 percent.[50] In several corridors connecting large European cities, airlines have discontinued service entirely following the introduction of high-speed rail.[51] In China, too, the emergence of high-speed rail led to the suspension of certain air routes, such as Nanjing-Wuhan.[52]

Intercity rail in Japan accounts for 18 percent of total domestic passenger-kilometers by all travel modes—compared with just 5–8 percent in major European countries and less than 1 percent in the United States.[53] In the United States, by contrast, air travel has 10 percent of total pkm volume, compared with Japan's 5 percent.[54] And in both the United States and Europe, private road vehicles account for an overwhelming 80-plus percent of domestic travel, whereas in Japan, cars have less than 50 percent.[55]

High-speed trains fare well in a comparison of environmental impacts. A 2006 tally of greenhouse gas emissions by travel mode done by the Center for Neighborhood Technologies found that HSR lines in Europe and Japan had emissions of 30–70 grams of carbon dioxide per passenger-kilometer (and conventional rail had about 60 grams per pkm).[56] That compared with 150 grams for automobile travel and 170 grams for airplane travel.[57] The figure of 150 grams actually understates the climate impact of cars. In 2006, cars in Japan averaged 149 grams, those in the European Union 160 grams, but those in the United States came in at 249 grams.[58] New regulations will bring down these values, but auto emissions will remain far higher than those of high-speed trains.[59]

In Japan, the Shinkansen connecting Tokyo and Osaka emits one twelfth as much carbon dioxide per passenger seat as air travel does (4.4 kilograms compared with 52 kilograms).[60] The newest Shinkansens use a third less energy than the original trains of the 1960s even though they travel considerably faster.[61]

Michael Renner is a senior researcher at World-watch Institute.

Table 3. Comparing Rail and Air Travel between Tokyo and Selected Japanese Cities

	Tokyo to Osaka	Tokyo to Hiroshima	Tokyo to Fukuoka
Distance (kilometers)	553	894	1,175
Rail travel time	2 hrs 25 min	3 hrs 48 min	4 hrs 51 min
Air travel time, airport to airport	1 hr	1 hr 15 min	1 hr 30 min
Air travel time, city center to city center	2 hrs 30 min	3 hrs 10 min	2 hrs 40 min
Rail departures per day	250	97	70
Air departures per day	102	30	92
Rail market share	82%	58%	10%
Aviation market share	18%	42%	90%

Source: Central Japan Railway Company, Data Book 2010 (Nagoya, Japan: 2010).

Environment and Climate Trends

A researcher at the Center for Advanced Energy Studies, Idaho, evaluates underground carbon storage methods

For additional environment and climate trends, go to vitalsigns.worldwatch.org.

Carbon Markets Struggle
to Maintain Momentum

Saya Kitasei and Will Bierbower

Between 2008 and 2009, the volume of global carbon transactions increased 80 percent, reaching 8.7 billion tons of carbon dioxide–equivalent (CO_2e).[1] (See Table 1.) Over the same period, declines in industrial output reduced demand for carbon assets, resulting in falling prices and an increase in the total value of global carbon transactions of only 6 percent.[2] The average global carbon price fell from $27.93 to $16.52 per ton of CO_2e.[3] (See Figures 1 and 2.)

Carbon markets (sometimes called emissions trading systems, or ETS) internalize the environmental costs of emitting carbon dioxide (CO_2) and other greenhouse gases (GHGs) by facilitating trade in allowances or permits for such emissions. This can be a mandatory or a voluntary system. In mandatory carbon markets, legislation or a binding international agreement establishes overall caps on the amount that countries, sectors, or specific segments of an industry may emit. Allowances to emit the established amount are allocated or auctioned to emitters, who surrender them when emissions

occur. Excess allowances may be sold to other emitters or, in some cases, banked for future use; some trading systems also permit emitters to purchase "offsets" or carbon reduction credits from projects outside the ETS. Voluntary carbon markets, in contrast, do not cap carbon emissions but they do allow carbon instruments, often offsets, to be traded among participants that may not be covered under a mandatory ETS. Carbon markets are now in use in over 30 countries.[4]

The European Union's Emissions Trading System (EU-ETS) remains the world's largest and most mature example of this, driving market trends and influencing market design and behavior around the globe. The EU-ETS completed its Phase I in 2007, which was intended as a trial period for market structure and operations. The Phase I cap was set to reduce emissions 4.3 percent below a business-as-usual scenario, and preliminary research estimates that the EU-ETS did reduce emissions 3–5 percent during Phase I.[5] In Phase II, which covers

Table 1. Volume of Carbon Transactions, Selected Markets, 2005–10

Market	2005	2006	2007	2008	2009
			(million tons of CO_2e)		
EU Emissions Trading Scheme	321	1104	2060	3039	6326
New South Wales	6	20	25	31	34
Chicago Climate Exchange	1	10	23	69	41
Regional Greenhouse Gas Initiative	0	0	0	62	805
Assigned Amount Units (AAUs)	0	0	0	23	155
Primary Clean Development Mechanism	341	537	552	404	211
Joint Implementation	11	16	41	25	26
Voluntary Markets	20	33	43	57	46
Total	710	1745	2984	4836	8700

Source: World Bank, State and Trends of the Carbon Market 2010 *(Washington, DC: May 2010).*

2008–12, the overall EU-wide market cap has been reduced to 6.5 percent below 2005 emissions levels and, as of 2010, covered about 11,000 medium-size and large emitters in industrial and energy sectors that are responsible for about 45 percent of the EU's CO_2 emissions.[6] In 2009, just over 6.3 billion tons of CO_2e were traded in the EU-ETS, which accounted for 73 percent of the global total of traded allowances, at a value of 88.7 billion euros ($118.5 billion).[7]

The European Commission opted not to allow emitters to bank Phase I allowances for use in Phase II due to concerns that the overallocation of allowances during Phase I would hinder the effectiveness of the Phase II cap. Nonetheless, the recession and resulting drop in industrial output and power generation led to a sharp decrease in demand for allowances, driving the prices of European Union Allowances (EUA) down to 8 euros per ton of CO_2e in February 2009.[8] Prices stabilized between 13 and 16 euros by May 2009 but so far have not recovered to their summer 2008 high of around 30 euros.[9]

Planning has already begun for Phase III of the EU-ETS, which will likely extend from 2013 to 2020. Major new features will include coverage of the aviation sector, a tighter cap (at least 21 percent below 2005 levels by 2020), a significantly larger proportion of allowances auctioned, and an extension of the trading system to cover greenhouse gases besides CO_2.[10] Formal linkage of the EU-ETS to other carbon markets has also begun: Iceland, Liechtenstein, and Norway joined the scheme in 2008.[11] Switzerland, also not an EU member state, is exploring the possibility of linking the domestic trading system it launched in 2008 to the EU-ETS in 2013.[12]

On the other side of the world, New Zealand launched the first mandatory economy-wide emissions trading system outside Europe in 2008, initially covering forestry and then adding the energy, industrial, and transport sectors in July 2010.[13] For the initial 2010–12 period, carbon prices are effectively capped at 12.50 New Zealand dollars (7 euros) per ton of CO_2e.[14]

Several other mandatory schemes are up

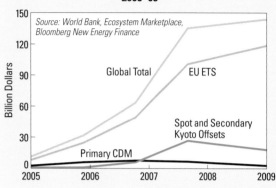

Figure 1. Transaction Values of Selected Carbon Markets, 2005–09

Source: World Bank, Ecosystem Marketplace, Bloomberg New Energy Finance

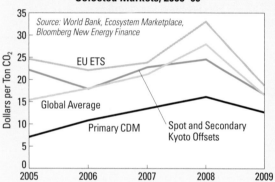

Figure 2. Weighted Average Annual Carbon Price, Selected Markets, 2005–09

Source: World Bank, Ecosystem Marketplace, Bloomberg New Energy Finance

and running at the sub-national level. The largest of these is the Regional Greenhouse Gas Initiative, which covers 95 percent of power sector emissions in 10 northeastern states in the United States.[15] In 2009, its first full year of operation, the volume of allowances traded rose to 805 million tons of CO_2e, although overallocation of allowances kept prices at an average of $3.30 per ton and total market value at $2.2 billion.[16] The market cooled considerably at the end of 2009 when prospects for a U.S. cap-and-trade bill diminished, the recession lowered electricity demand, and low natural gas prices reduced the market share of carbon-intensive coal-fired electricity.[17]

The Canadian province of Alberta, which launched an emissions trading scheme in 2007 covering its highest industrial emitters, saw carbon prices rise from $10 in 2008 to about $13.50 in 2009, while traded volumes increased from 3.4 million to 4.5 million tons of CO_2e.[18] These prices are close to the trading system's effective cap of 15 Canadian dollars per ton of CO_2e, which companies in the carbon market can choose to pay instead of purchasing and surrendering allowances.[19] The Australian state of New South Wales has been operating a mandatory emissions trading system for electricity producers since 2003, with producers required to purchase certificates from emissions reduction projects.[20] In 2009, permits representing 34 million tons of CO_2e were traded for $117 million on the New South Wales ETS.[21]

Several more emissions trading systems are under active development. California's scheme is set to begin in 2012.[22] China, which became the world's highest CO_2 emitter in 2006, has planned to set "binding" carbon emissions targets on energy and GHG-intensive sectors in the twelfth Five-Year Period (2011–15) in order to meet its 2020 carbon emissions intensity targets.[23] The country is currently studying various policy mechanisms to establish a preliminary ETS.[24]

Elsewhere in the world, however, several countries put plans to implement emissions trading systems on hold after climate talks in December 2009 failed to produce an internationally binding treaty to reduce GHG emissions. The U.S. Senate lost momentum on a cap-and-trade bill that passed the House of Representatives in the summer of 2009, and the results of midterm elections in 2010 make the passage of a climate bill during the 112th Congress highly unlikely.[25] Similarly, a bill enacting the Australian Carbon Pollution Reduction Scheme passed the lower house of Australia's parliament in 2009 but stalled in the Senate.[26] And strong opposition from industry has led Japan to pull back from plans to implement a mandatory ETS in 2013.[27] The South Korean

government's commitment to a trading system that would launch in 2013 is likewise wavering.[28] Even New Zealand may not follow through on the next phase of its ETS unless there is sufficient evidence of commitments to carbon reduction elsewhere in the world.[29]

In some markets, the waning appetite for emissions trading schemes may provide an opening for carbon taxes or tariffs. Japan is still planning to implement a carbon tax on fossil fuels.[30] Mounting doubts in the EU about the ability of other major emitters to restrict their GHG emissions have sparked arguments for the use of carbon border tariffs, which would in effect create a "shadow" carbon price for industries outside the EU and thereby prevent companies from leaving Europe to avoid carbon prices, a phenomenon called "carbon leakage."[31]

Falling demand for allowances in the EU-ETS and other trading schemes reduced demand for offsets under the Kyoto Protocol. Countries mandated to meet emissions reduction targets under this treaty (known as Annex B countries) are given a specified number of credits, called Assigned Amount Units (AAUs), representing the amount of CO_2e they are permitted to emit during the compliance period of 2008–12. The Kyoto Protocol also instituted the Clean Development Mechanism (CDM), under which Annex B parties can purchase emissions offsets from projects in non-Annex B countries, and the Joint Implementation (JI) program, under which Annex B countries can purchase emissions offsets from projects in other Annex B countries.[32]

As financing for CDM projects declined over 2008 and 2009, the demand quota for offset credits has been met primarily by AAUs from governments that are ahead of schedule for meeting emissions reduction targets, such as the Czech Republic and Ukraine.[33] Transaction volume on the AAU market increased almost 700 percent from 2008 to 2009, from 23 million to 155 million tons of CO_2e.[34]

In 2009, the volumes traded under the CDM fell by 48 percent to 211 million tons of CO_2e, while the average price of CDM offsets—certified emissions reductions—fell 21 percent to $12.70 per ton from $16.10 per ton in 2008.[35] Transaction volume fell for the second year in a row from a high of 593 million tons CO_2e.[36] China remained the largest seller of CDM offsets, contributing 72 percent of the global market.[37] Transactions on the JI market increased slightly to 26.5 million tons in 2009, but the average price fell 8 percent to $13.40 per ton.[38] The outlook for the CDM and JI markets remains uncertain, as the Kyoto Protocol's first commitment period ends in 2012 but there is no comprehensive international plan to replace it.

The global economic recession hit voluntary carbon markets especially hard. The volume of 2009 voluntary market transactions declined 26 percent from 2008, although it remained 39 percent above the 2007 level.[39] The prospect of U.S. climate legislation buoyed markets in 2009 as actors bought offset credits in the hopes that they would become compliant in a U.S. emissions trading scheme. Most offsets generated in the voluntary market come from the energy sector in the form of landfill methane capture and renewable energy projects, but a significant and increasing portion come from forestry projects.[40] Interest is growing in the international policy community to develop strategies and projects for reducing emissions from deforestation, and offsets from these projects are expected to enter into the compliance markets first with California's emissions trading scheme in 2012.[41]

Saya Kitasei was a MAP Sustainable Energy Fellow and Will Bierbower was a Climate and Energy research intern at Worldwatch Institute.

Carbon Capture and Storage Attracts Government Attention

Saya Kitasei and Matthias Kimmel

Between 2005 and 2009, some $25 billion of public investment was announced in support of carbon capture and storage (CCS) projects.[1] (See Figure 1.) In 2010, new announcements of CCS investments declined, however, as governments shifted from announcing funding, especially as part of stimulus packages, to allocating funds to specific projects.[2]

As of March 2011, the Global CCS Institute had identified a total of 79 large-scale, fully integrated CCS projects in 17 countries at various stages of development, but only 8 of these were operational.[3] (See Figures 2 and 3.) These 8 projects store a combined total of 11.08 million tons of carbon dioxide (CO_2) per year (Mtpa), equivalent to the amount emitted annually by 2.2 million passenger vehicles in the United States .[4] If the remaining 71 projects under planning or development are built, they would add an estimated 152 Mtpa of capacity.[5]

Governments and industry have been invest-ing heavily in CCS for several years with the aim of substantially decreasing CO_2 emissions from the fossil-fueled power sector, especially from greenhouse gas–intensive coal plants, although CCS can also be used in natural gas power plants and a range of industrial facilities. CCS can cut CO_2 emissions in coal-fired power plants by 85–95 percent, an essential modification if coal is to continue to provide substantial amounts of power to countries committed to lowering greenhouse gas emissions.[6]

Some 77 percent of government funding for CCS since 2005 has targeted power generation projects.[7] Although about 95 Mtpa of CCS capacity (45 projects) for large-scale power plant projects is currently under development—with small pilot projects in Australia, China, France, Germany, Italy, the Netherlands, Norway, Sweden, and the United States—no large-scale commercial power plants with CCS were in operation as of the end of 2010.[8]

CCS involves a three-step process: isolating CO_2 from a source such as natural gas or a power plant's flue gas, transporting this CO_2 to a storage site, and injecting it into the storage reservoir.[9] There are three primary methods for CO_2 capture in power plants: post-combustion, pre-combustion, and oxy-fuel combustion.[10] Thus far, no single approach has emerged as the clear winner in terms of cost or feasibility. Pre-combustion and post-combustion technologies have attracted almost the same amount of investment ($3.5 billion and $3.6 billion respectively). Investments in oxy-fuel CSS are significantly smaller ($1.6 billion).[11]

In post-combustion, CO_2 is extracted from flue gases that emerge from the combustion process. Because it requires little modification to combustion technology, it has received broad attention from the industry. Twenty-five post-

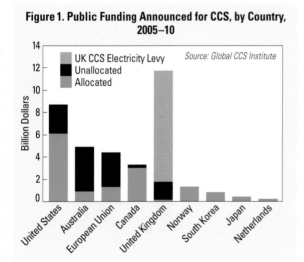

Figure 1. Public Funding Announced for CCS, by Country, 2005–10

Source: Global CCS Institute

Legend: UK CCS Electricity Levy; Unallocated; Allocated

combustion CCS projects are currently under development, though this method has not yet been demonstrated in a large-scale integrated CCS project.[12]

Pre-combustion combines gasification of a typically solid fuel with CO_2 separation to yield a hydrogen gas, which can then be burned without emitting greenhouse gases. Although pre-combustion has not yet been demonstrated in a power plant pilot project, two operational projects use this technology: the Weyburn Great Plains Synfuels plant and the Enid Fertilizer plant. An additional 36 projects are currently under development.[13]

Oxy-fuel technology combusts fuel in oxygen mixed with recycled flue gas rather than nitrogen-rich air, producing a storage-ready CO_2-rich gas. This technique has been demonstrated with industrial processes such as steel manufacturing.[14] Five oxy-fuel projects, in Germany, Spain, the United Kingdom, and the United States, are in the very early stages of conceptualization.[15]

Natural gas processing is a fourth, established method of carbon capture that has been in use for almost four decades.[16] During gas processing, CO_2 that occurs naturally in recovered gas is extracted to meet the commercial specifications of natural gas. Historically, the resulting relatively pure CO_2 stream has typically been vented, but six natural gas processing facilities currently capture the CO_2 for transport and storage, and an additional six large-scale integrated CCS projects using natural gas processing are currently under development.[17]

CO_2 has been transported, in some cases over hundreds of kilometers, for decades in North America and, to a much smaller extent, Europe.[18] For the most part CO_2 is transported to storage sites in pipelines, but vehicles may also be used in regions where pipeline transport is not feasible.[19]

Three main options currently exist for carbon storage: saline aquifers, depleted oil and gas reservoirs, and deep unminable coal seams.[20] In terms of global capacity, saline formations have the greatest potential. Three natural gas processing facilities—Sleipner, Snøhvit, and In Salah—

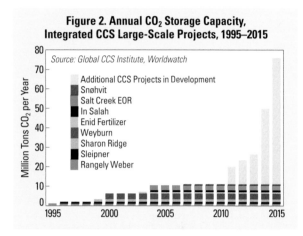

Figure 2. Annual CO_2 Storage Capacity, Integrated CCS Large-Scale Projects, 1995–2015

Source: Global CCS Institute, Worldwatch

- Additional CCS Projects in Development
- Snøhvit
- Salt Creek EOR
- In Salah
- Enid Fertilizer
- Weyburn
- Sharon Ridge
- Sleipner
- Rangely Weber

Figure 3. Global Planned and Active CCS Capacity, by Country or Region, 2010

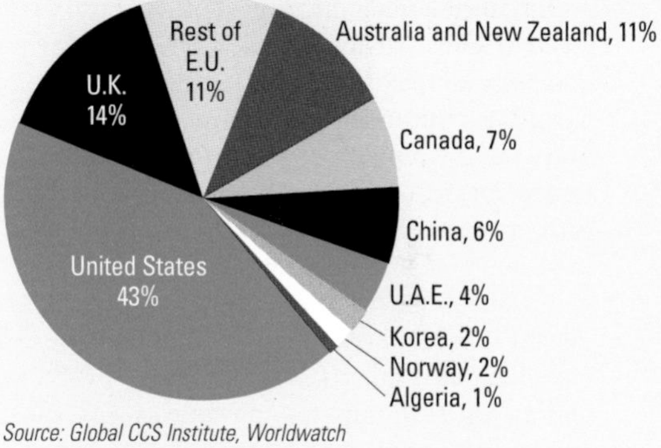

Australia and New Zealand, 11%
Rest of E.U. 11%
U.K. 14%
Canada, 7%
China, 6%
U.A.E., 4%
United States 43%
Korea, 2%
Norway, 2%
Algeria, 1%

Source: Global CCS Institute, Worldwatch

store CO_2 from their operations in nearby saline geological formations.[21]

Thus far, oil reservoirs have received the greatest investment for carbon storage. Injection of captured CO_2 into oil wells allows producers to extract economically inaccessible oil and significantly extend the life cycle of an oil field, a process known as enhanced oil recovery (EOR). Five existing projects inject CO_2 into depleted oil reservoirs to enhance oil recovery.[22] Although EOR with CO_2 from human activities still comes at a cost and cannot be applied everywhere, its use has become economically viable since oil and gas prices have risen.

The incremental cost of including CCS in a power plant remains prohibitively high. In 2011, the International Energy Agency published a working paper on the costs of CCS in power plants in countries that belong to the Organisation for Economic Co-operation and Development (OECD); it found that adding CCS increased the levelized cost of electricity by 33 percent for natural gas plants, leading to a price of 10.2¢ per kilowatt-hour (kWh). For coal, electricity costs increase between 39 and 64 percent with CCS, resulting in prices between 10.2¢ and 10.7¢/kWh, depending on which CCS technology is used. Using carbon capture technology also consumes some of the electricity generated by the power plant, causing relative plant efficiency losses of 15 percent in natural gas power plants and 20–25 percent in coal-fired plants.[23]

Environmental concerns about the use of CCS are vital, particularly as the technology is being promoted as a way to maintain the use of fossil fuels for power generation as economies move away from carbon emissions. Industry representatives and some scientists claim that carbon can be stored for hundreds of thousands of years.[24] Others, however, suggest that CCS will probably never be completely leak-free. Recent reports of carbonation in pond water and blowouts on residents' land have given CCS opponents fresh impetus.[25] While at least one study claimed this was due to a failure of CO_2 storage at the nearby sequestration site, other researchers maintain that the observed phenomena were not linked to the Weyburn Great Plains Synfuel plant.[26] In addition, CCS significantly increases water usage and requires more consumption of fossil fuels to balance efficiency loss at power plants, and it can lead to drinking water contamination.[27] As governments plow ahead with support for CCS, addressing these concerns will become increasingly important.

Some progress has been made during the last year in developing CCS regulations. In recent years, some OECD countries, including Australia, Canada, Japan, South Africa, South Korea, and the United States, have introduced new or improved legal frameworks that address the different challenges and opportunities of CCS, including environmental risks, long-term liability, storage reservoir ownership, financial contribution to post-closure stewardship, and CCS readiness. Non-OECD countries, including major CO_2 emitter China, are lagging behind in CCS regulatory development.[28]

Although CCS has received increased attention on the international stage, implementation of an international regulatory framework is progressing slowly. In November 2006 and October 2009, the Convention on the Prevention of Marine Pollution by Dumping of Wastes and Other Matter (the London Protocol) was amended to allow for offshore CO_2 storage and cross-border transportation. The latter, however, requires mandatory ratification, which had not made much progress as of early 2011. International negotiators on the climate change treaty, after five years of stalled negotiations on this topic, agreed at the 16th Conference of Parties in Cancun, Mexico, that CCS is eligible under the Clean Development Mechanism, but procedures to include it may take some time to develop.

Saya Kitasei was a MAP Sustainable Energy Fellow and Matthias Kimmel was a Climate and Energy research intern at Worldwatch Institute.

Food and Agriculture Trends

Rice varieties on sale at a market in Kandy, Sri Lanka

For additional food and agriculture trends, go to vitalsigns.worldwatch.org.

World Grain Production Down in 2010, but Recovering

Richard H. Weil

Maize, rice, and wheat are the three major grain crops in the world. Together they account for almost two thirds of humanity's staple food intake.[1] In addition, they are critical for animal feed and industrial uses. In 2010, production of both maize and rice set record levels, but a significant drop in wheat output left overall grain totals slightly below 2008 levels.[2] (See Figure 1.) Preliminary data for 2011 indicate that production increased, however, and the U.N. Food and Agriculture Organization (FAO) recently forecast that cereal output in 2011–12 will be 3 percent higher than in 2010–11.[3]

Since the 1960s maize production has quadrupled, while paddy rice and wheat output have tripled.[4] From 1961 to 2010 the world's annual harvest of these three crops increased from 643 million to 2.2 billion tons.[5] This is remarkable, given that the amount of land used for agriculture increased only 35 percent.[6] (See Figure 2.) Thus the average farmland yield increased by 156 percent.[7] (See Figure 3.)

Much of this growth occurred because of the Green Revolution, in which high-yield crop varieties were widely introduced from 1960 to 1990. Production increased worldwide, but there was greater reliance on irrigation, synthetic fertilizers, and pesticides—all of which take resources, can be costly, and may cause substantial environmental degradation.[8] In addition, there are limits to the levels of hybrid crops that can be developed or deployed for use. Because of these factors, in the late 1980s the rate of growth began to decline.[9] With the exception of a recent upturn in maize, this pattern has continued.[10] (See Table 1.)

Despite the large production gains of the last half-century, the increases in yield look less impressive when population growth is included. Between 1961 and 2009 the total production of maize, rice, and wheat increased by 241 percent, but on a per capita basis it was up 54 percent.[11]

In addition, the growth in output has not occurred evenly around the world. In many poor countries, food supplies remain expensive and scarce, which is reflected in consumption patterns. While during 1997–99 the world's daily average per capita consumption of food equaled 2,803 kcal, in industrial countries the figure was 3,380 kcal compared with 2,681 kcal in developing countries.[12]

In 2010, world wheat production dropped 5 percent.[13] The major factor was a severe drought in Russia and northern Kazakhstan, extending from the Volga River to the Urals. Russia's overall wheat yield declined 37 percent and the country banned export of the crop.[14] This had a major effect on world markets, given that Russia was the fourth largest wheat exporter in 2009, selling 18 million tons.[15] At the same time, unstable weather lowered wheat production in Europe, and flooding reduced Canadian output by 17 percent.[16]

Poor weather in the Ukraine also reduced the

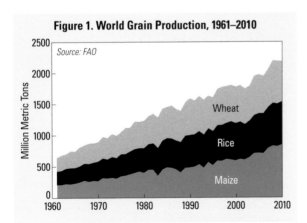

Figure 1. World Grain Production, 1961–2010

Source: FAO

62

maize crop in 2010, although world output increased.[17] U.S. maize production was down 5 percent due to drought in the East and excessive rain in the West.[18] This was significant, since the United States is by far the largest exporter of maize in the world—50 million tons in 2010, which was 54 percent of the global total.[19]

Exports may face some limits in the future due to increasing internal demand.[20] While U.S. domestic demand for maize for animal feed has been relatively steady (28 percent of world total), since the 1990s there has been significant growth in the demand for ethanol.[21] According to the Congressional Budget Office, about 20 percent of the increase in maize prices between 2007 and 2008 was due to domestic ethanol demand.[22] And the International Food Policy Research Institute maintains that 40 percent of the increase in maize prices in 2000–07 was due to global demand for ethanol.[23]

Rice production in 2009 was affected by an El Niño event, when a warming of the waters of the western Pacific disrupted the usual atmospheric circulation.[24] As a result, rice output fell across Southeast Asia. Production rebounded in 2010, however; improving weather conditions in India was one factor in a 2 percent increase in world rice production, thereby setting a new record.[25] Despite rising production, higher world demand kept prices high.[26] Price policies in Thailand, the largest rice exporter, contributed to upward price pressure.[27]

FAO tracks the cost of wheat, rice, and maize in the Cereal Price Index. Using 2002–04 prices as the baseline of 100, by August 2010 the index had reached 185 and world food prices increased 5 percent in the month after Russia embargoed export of its wheat.[28] In April 2011, the Cereal Price Index set a record at 265.[29] There were many reasons for this rise, including a reduced maize crop in the United States, the weakening of the dollar, speculation, demand for industrial and biofuels uses, and possible climate change impacts.[30]

In August 2011, the index dropped back down to 253 as better weather led to increased world wheat and rice production and the end of

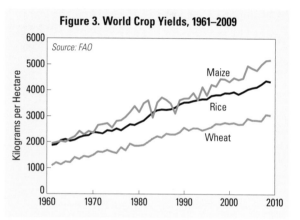

Figure 2. World Grain Harvest Area, 1961–2009

Source: FAO

Figure 3. World Crop Yields, 1961–2009

Source: FAO

Table 1. Increase in Worldwide Production of Maize, Rice, and Wheat, 1971–2010

Crop	1971–80	1981–90	1991–2000	2001–10
		(percent)		
Maize	47.8	27.0	24.7	30.4
Rice	33.2	32.4	20.0	13.6
Wheat	39.6	31.2	12.1	8.9

Source: FAO, FAOSTAT Statistical Database.

Russia's wheat export ban.[31] Yet this was still a gain of almost 37 percent in one year. It is likely that this trend will continue and that tighter supplies and higher prices can be expected to become the norm for all grains.[32]

Farming has always been an uncertain business dependent on the weather, and it could be entering an even more difficult phase. The warmer, less-stable atmosphere as the climate changes due to human activities could be detrimental for food production.[33] One computer model examined world temperature and precipitation trends from 1980 to 2008 and translated them into potential yields, showing noticeable changes in grain production.[34] While, on the plus side, rice output may have slightly increased (2.9 percent), for wheat and maize the model showed global losses of 3.8 and 2.5 percent, respectively.[35] The biggest effect was found in Russia, where up to 15 percent of the wheat output during these years may have been lost due to human-induced drought.[36] If accurate, the model implies continuing and significant price increases for these crops.

Other factors to consider are that demand is rising as the middle class grows in a number of countries, notably China and India. At the same time, the population in the poorest nations continues to increase; even at lower standards of living, these people need more food.

In a time of economic slowdown, higher prices and unpredictable weather-induced shortages will continue to have severe negative effects on low-income countries. As the executive director of the World Food Programme, Josette Sheeran, noted, we now live in a "post-surplus world."[37] Reaching a sustainable level of food production remains a challenge.

Richard H. Weil teaches general education at Brown College in Mendota Heights, Minnesota. He would like to thank Lindsay Nauen for reviewing earlier drafts.

Organic Agriculture Sustained through Economic Crisis

E. L. Beck

In 2009, organic farming was practiced on 37.2 million hectares worldwide, a 5.7 percent increase from 2008 and a 150 percent increase since 2000.[1] (See Figure 1.) This includes land that is transitioning to organic production. The organic area amounted to 0.85 percent of global agricultural land in 2009.[2] (By comparison, producers seeded 2 percent of agricultural land worldwide with genetically modified crops.)[3]

Although the term "organic agriculture" has many meanings, the International Federation of Organic Agriculture Movements defines it as follows: "Organic agriculture is a production system that sustains the health of soils, ecosystems and people. It relies on ecological processes, biodiversity and cycles adapted to local conditions, rather than the use of inputs with adverse effects. Organic agriculture combines tradition, innovation and science to benefit the shared environment."[4]

Organic agriculture is practiced worldwide, but certified organic agriculture tends to be concentrated in wealthier countries. (See Table 1.)

Countries belonging to the Group of 20, which includes both industrial and developing economies, have 89 percent of the global certified organic agricultural area.[5] (See Figure 2.) Out of 1.8 million organic producers worldwide, India leads with just over 677,000, followed by Uganda with nearly 188,000 and Mexico with nearly 129,000.[6]

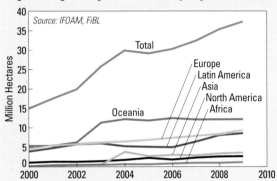

Figure 1. Organic Agricultural Land by Region, 2000–2009

Source: IFOAM, FiBL

Total
Europe
Latin America
Asia
North America
Africa
Oceania

Million Hectares

2000 2002 2004 2006 2008 2010

Table 1: Organic Agricultural Land, by Continent, 2007 and 2009

Region	Organic Area, 2007	Organic Area, 2009	Increase, 2007 to 2009	Share of Area's Agricultural Land
	(hectares)			(percent)
Africa	862,351	1,026,632	19.0	0.10
Asia	2,895,763	3,581,918	23.7	0.25
Europe	7,769,157	9,262,997	19.2	1.87
Latin America	6,414,709	8,558,910	33.4	1.37
North America*	2,292,357	2,652,624	15.7	0.68
Oceania	12,074,550	12,152,108	0.6	2.82
Total	32,308,887	37,235,189	18.6	0.85

*Does not include U.S. data for 2009.
Source: Helga Willer and Lukas Kilcher, eds., The World of Organic Agriculture: Statistics and Emerging Trends 2011 (IFOAM and Research Institute of Organic Agriculture, 2011).

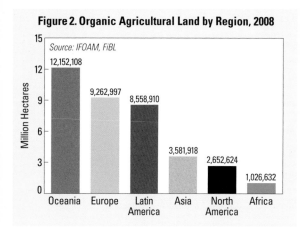

Figure 2. Organic Agricultural Land by Region, 2008

Source: IFOAM, FiBL

Million Hectares

Oceania 12,152,108
Europe 9,262,997
Latin America 8,558,910
Asia 3,581,918
North America 2,652,624
Africa 1,026,632

Europe's organic farmland reached 9.26 million hectares in 2009.[7] The newest EU members report the fastest growth in organic farmland. Emerging as a center for organic research, European nations funded research programs, organic action plans, and Europe-wide projects, all of which have contributed to the growth of organic agriculture.[8]

For North America, organic agricultural land covered 2.65 million hectares (although U.S. data are through 2008 only).[9] Organic farmland in the United States and Canada represents 0.68 percent of the region's agricultural land, below the global average of 0.85 percent.[10] Imports compensate for the dearth of organic production in North America. In 2010, the Canadian government funded organic research to improve production, putting $6.5 million of its own money into various projects along with $2.2 million from the U.S. organic industry.[11]

Organic production in other regions is driven by exports, primarily to Europe and North America.[12] Certified organic agricultural land in Africa has increased to 1.03 million hectares.[13] Coffee and olives are the principal organic crops there, most of which is exported, particularly to the European Union.[14] The International Assessment of Agricultural Knowledge, Science and Technology for Development promoted organic techniques to revive idled or barren land in Africa.[15] For example,

organic agroforestry shows promise in rejuvenating desert land.[16] Over two decades, this technique has rehabilitated some 350,000 hectares of western Tanzanian desert. Similar techniques using "fertilizer trees" (indigenous varieties whose roots set nitrogen into the soil) are being used in Malawi.[17]

Ethiopian farmers are working with non-governmental organizations—including Slow Food International, ACDI/VOCA (a group that works with farmers and business throughout Africa), and the Ethiopian Coffee Exchange—to learn how to protect wild coffee plants, fertilize them with organic compost, and process them to retain the qualities savored by coffee drinkers.[18]

Latin America's organic farmland grew to 8.56 million hectares in 2009, a 33.4 percent increase in two years.[19] Some organic farmland areas in the region have grown faster than others.[20] For instance, Argentina saw a 53.3 percent increase in certified organic land, now second in the world, behind Australia but ahead of the United States.[21] Argentina's growth has been assisted by its preferential trade status with the EU.[22] Brazil and Colombia have applied for this status, which could have the same impact on organic sector growth there.[23] Latin America exports the majority of its organic products to Europe, North America, and Japan.[24] International traders, foreign investors, and governments are stimulating export projects, but size and scale places them beyond the means of local companies and producers.[25]

Several issues hinder further expansion of organic agriculture in Latin America. Bird-friendly and fair-trade agricultural standards, as well as standards set by the Rainforest Alliance, compete with organic standards. Some of the non-organic standards allow the use of synthetic inputs, require fewer production changes, and do not involve organic certification costs—all of which increase the chance that farmers will adopt them. In addition, prices received by producers for organic products are not always higher than those of fair trade or other alternative market products.[26] Another limiting factor for organics is pest and fungus control, which is of particular concern in the region and which can be difficult using organic methods.[27] This underscores the need for research unique to each region's demands.

In Asia, organically managed land reached 3.58 million hectares in 2009.[28] In Sri Lanka and Indonesia, there is growing support for organic fertilizers as an alternative to expensive synthetic fertilizers, thus reducing input costs.[29] Across Oceania, 12.15 million hectares of agricultural land were organically managed as of 2009—the largest amount of any continent.[30] Nearly all of this land is in Australia.[31] Between 2002 and 2003, the region experienced an 80.7 percent increase in organic agricultural land, but since then the area has grown just incrementally.[32]

The global organic market is recovering from the recession. In 2009, growth slowed to single digits, 5 percent, for the first time since 2000.[33] Global organic sales rose to $54.9 billion in 2009.[34] Demand is concentrated in Europe and North America, which each account for 48 percent of global sales.[35] Germany, the United Kingdom, France, and Italy represent over 70 percent of EU organic sales.[36] Although growth in the European Union's organic market slowed, individual country trends differed. In the United Kingdom, the organic market declined nearly 13 percent during 2009, and in Germany it stagnated that year. Yet sales of organics continued to grow by double digits in France, Switzerland, and Sweden.[37] Some of the slowdown in growth in Europe's sales of organic products can be attributed to prices declining while sales volumes rose.[38]

In 2009, organic retail sales in the United States climbed to $26.3 billion, a 5.3 percent increase from 2008.[39] Yet this represents a slower pace, as annual double-digit increases had become the norm since 2000.[40] For U.S. organic food products, 82.6 percent of first-time sales were made to wholesalers and just 10.6 percent were sold directly to consumers.[41] There is also a shift occurring in retail outlets: Large retailers (such as Wal-Mart and Krogers) accounted for 54 percent of total U.S. organic food sales in 2009, up from 45 percent in 2008,

while organic retailers (such as Whole Foods and Trader Joe's) sold 38 percent, down from 43 percent a year earlier.[42]

Organic price premiums can be a barrier to wider adoption of organic foods.[43] Some consumers see organics as expensive, despite the environmental, ethical, and potential health benefits and the current media coverage about the hidden costs of conventional foods.[44]

Two challenges within the global organic movement are the lack of organic standards within some countries and the scarcity of equivalency agreements.[45] An equivalency agreement between two nations acknowledges each other's organic standards as acceptable for labeling purposes. The Global Organic Market Access (GOMA) project, financed by the Norwegian Agency for Development Cooperation, facilitates trade in organic products by harmonizing organic standards and generating equivalency agreements.[46]

Asia's certification process is one example of the need for the GOMA project. China, India, Japan, South Korea, the Philippines, Taiwan, and Malaysia each recently developed organic certification standards. While this has facilitated trade with North America and Europe, it has created obstacles to trade within Asia due to the differing standards.[47] Small Asian producers also find it difficult to stay abreast of evolving regulations.[48] One GOMA-like solution to the problem in Asia is the Pacific Organic Standard (POS). The International Organic Accreditation Service assessed the POS in 2010 and found it to meet the equivalency of the EU regulations. Now Pacific region certification bodies can use the POS as the standard for certifying organic producers wanting to export to Europe.[49]

While international organic standards assure consumers, they can also create problems. In the South Pacific, traditional farming techniques use sustainable methods free from synthetic inputs, thus retaining the essential characteristics of organic farming. As the impetus has shifted from subsistence farming to exporting organic products, however, farmers must now bear the higher costs of certification, auditing, and compliance for agricultural methods that have not changed.[50] In addition, rising fuel and transportation costs may decrease the economic viability of exporting organic products in the future.

Farmland prices around the globe are rising, with ramifications for organic agriculture.[51] Two factors driving this trend include investors searching for more-stable investment opportunities and governments seeking food security elsewhere for their own countries.[52] While higher prices may not affect conventional agriculture or large-scale organic operations (such as Latin American export projects), they will make it harder for small producers to enter organic agriculture or expand existing operations, hence slowing the conversion to organic farmland. In frontier and developing economies, particularly in Africa, rising farmland values threaten indigenous farmers if governments do not recognize

tenure agreements and sell or lease the sought-after land to foreign investors.[53] In industrial economies, higher initial capital costs caused by rising land prices cannot be recovered with increased product prices in light of consumer hesitancy to pay more for organic produce.[54] These situations represent potential barriers to the continued growth of local organic food systems. Despite these challenges, however, organic agriculture holds untapped potential for helping farmers and consumers alike build resilience to food price shocks, climate change, and water scarcity.

E. L. Beck is an independent policy analyst in the U.S. Midwest who has used sustainable farming methods to revive depleted topsoils and worked on the local food movement in Chicago and on community gardens.

Sugar Production Dips

Jorge Moncayo and Gary Gardner

Global production of sugar crops declined by 2.4 percent in 2009 to approximately 1.9 billion tons.[1] (See Figure 1.) The drop follows three consecutive years of sharp increases: between 2005 and 2008, sugar production had increased by an average 7.5 percent annually.[2] The 2009 decline is related to the global recession and to a steep increase in the price of sugar.

Sugar crops, as defined here, include sugarcane and sugar beets. Other sweeteners such as honey, maple syrup, and artificial sweeteners are not included. Sugar crops were once largely used for food, but they are increasingly used to make ethanol for automobiles.

Sugarcane has grown in importance relative to sugar beets over the last half-century. (See Table 1.) Some 1.7 billion tons of sugarcane were produced worldwide in 2009, nearly quadruple the output of 1961, while harvested area increased by 2.6 times and yields by 41 percent.[3] Sugar beet production, in contrast, increased by just 43 percent, and the harvested area shrank by 38 percent.[4] Today, sugarcane accounts for about 88 percent of global sugar crop production.[5]

Although the U.N. Food and Agriculture Organization lists more than 100 nations that grow sugarcane, production is highly concentrated, with Brazil, India, and China accounting for more than half of global output (36, 15, and 6 percent, respectively).[6] Brazil has seen the sharpest growth over the past decade, while China's output has increased moderately and India's production has been volatile. (See Figure 2.)

The increase in Brazilian production is to some extent due to expanded area: Brazil more than doubled its cane area in the past two decades, to nearly 7.5 million hectares.[7] This is in part due to a growing international market in which Brazil is very competitive: its production

costs are some 25 percent lower than the global average.[8] In addition, Brazil has captured global market share as other suppliers have left the market. The European Union, for example, which accounted for as much as 20 percent of global sugar exports in the 1990s, is now a net importer because of sugar policy changes enacted in 2005.[9] Cuba was a major producer until the early 1990s, when the country diversified its agriculture beyond sugar; output in 2009 was only 18 percent of its 1990 level.[10]

China and India saw declines in output in

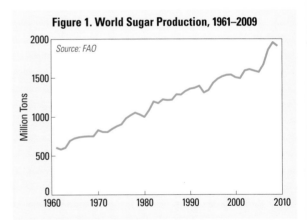

Figure 1. World Sugar Production, 1961–2009

Source: FAO

Table 1. Performance of Sugarcane and Beet Sugar Crops, 1961–2009

	Percentage Change in		
	Production	Harvested Area	Yield
Sugarcane	276	166	41
Sugar beets	43	−38	129

Source: FAO, FAOSTAT Statistical Database.

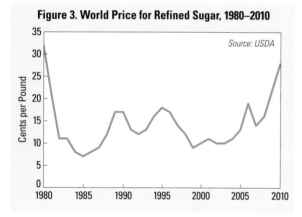

Figure 2. Sugar Crop Production, Selected Countries, 1961–2009

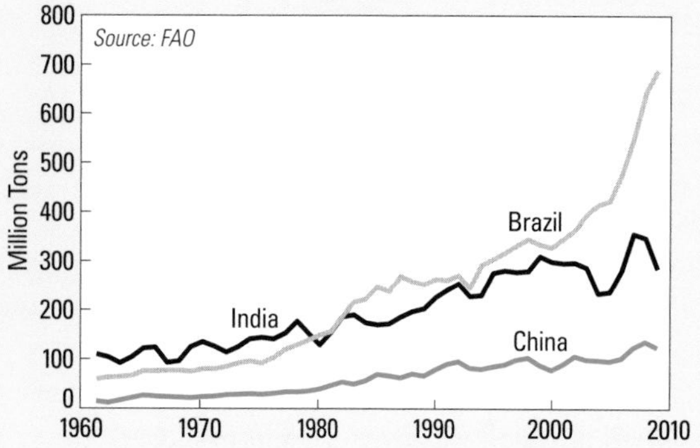

Source: FAO

Million Tons

Brazil

India

China

1960   1970   1980   1990   2000   2010

Figure 3. World Price for Refined Sugar, 1980–2010

Source: USDA

Cents per Pound

1980   1985   1990   1995   2000   2005   2010

2009. Sugar crop production in China decreased by some 8.6 percent, although 2009 production was still some 29 percent ahead of the 2006 level.[11] India's sugar crop production fell by 18 percent in 2009, but remained 22 percent ahead of its 2004 level.[12] In India's case, volatility in production is the norm: the lowest output over the past decade was less than half the level of the highest output.[13] Swings in Indian production have shifted the country from being a net exporter to a net importer and back again.[14] Indian policy is often the cause: state and national governments have policies affecting sugar storage, pricing, and trade.[15]

Policy-driven production volatility in India, China, and other Asian nations, combined with poor weather in some sugar-growing regions, helped to push world sugar prices to nearly 30¢ a pound in 2010 (see Figure 3)—a 29-year high, and about 50 percent above the average price level of the past two decades.[16] Long-term structural market shifts were also in play. The global sugar market is changing as sugar plays a greater role in energy production and as the exchange rate for Brazilian currency changes.

One of the structural shifts driving sugar prices up, as noted earlier, is the rapidly expanding global demand for biofuels. Ethanol production (from both corn and sugar) increased by nearly 10 percent in 2009, and by nearly 44 percent the year before that, driving up demand for sugar.[17] Indeed, increased demand for biofuels over the past decade has been largely responsible for the increase in sugar area in Brazil.[18] Today, some 55–60 percent of Brazil's sugar crop goes to ethanol production.[19]

Sugar production has a number of environmental impacts, including deforestation and habitat loss, water pollution and eutrophication, soil erosion and soil compaction, air pollution, and water scarcity.[20] Much of the impact depends on local circumstances, such as whether production opens up virgin lands or is done on existing agricultural land and whether water is sustainably available for sugar production and processing.

A multistakeholder sugar group known as Bonsucro (formerly the Better Sugar Initiative) was formed by sugar producers, traders, and nongovernmental organizations to reduce the environmental and social impact of sugar production. In December 2010, the group launched the Bonsucro Standard, a set of metrics designed to advance its sustainability objectives. Among other things, the standard requires stakeholders to comply with International Labour Organization conventions governing child labor, forced labor, and the right to collective bargaining; to pay at least the prevailing minimum wage; to evaluate the impacts of sugarcane enterprises on biodiversity and ecosystem services, and to conduct an environmental and social impact assessment of any new sugar cultivation.[21]

Jorge Moncayo was an intern and Gary Gardner was a senior researcher at Worldwatch Institute.

Fish Production from Aquaculture Rises While Marine Fish Stocks Continue to Decline

Matt Styslinger

Global fisheries production from fish caught in the wild and from aquaculture (fish farming) reached 145.1 million tons in 2009 (the most recent year with data), an all-time high.[1] (See Figure 1.) This represents a 1.9 percent increase from 2008, slightly higher than the previous year's 1.8-percent growth rate.[2] Forecasts for 2010 suggested a growth of 1.3 percent to 147 million tons.[3]

Both wild capture and aquaculture are practiced in marine and freshwater environments. (See Figure 2.) Wild capture output increased by 0.22 percent from 2008 to 2009 and accounted for roughly 60 percent of all fish production.[4] Aquaculture output, in contrast, grew approximately 4.5 percent in 2009.[5] Once a minor contributor to total fish harvest, aquaculture production increased some 50-fold to 52.6 million tons between the 1950s and 2008, and it is set to surpass output of wild fisheries within a few years.[6]

Capture fish production dominated marine fisheries, accounting for 80 percent of the 99.2 million tons of marine fish harvested in 2008.[7]

In freshwater (inland) fisheries, however, production was dominated by aquaculture, which accounted for just over 76 percent of the 43.1 million tons of fish produced.[8] Thus inland aquaculture accounted for approximately 88 percent of the total growth in global fish production—both marine and inland—in 2008.[9]

Global fish captures peaked in 1996 at 86.3 million tons.[10] Although capture production has leveled off since then, many fishery stocks have remained in decline.[11] About 32 percent of global marine fishery stocks were considered overexploited, depleted, or recovering in 2008.[12] This is 7 percent more than just three years earlier.[13] And in 2008, about 53 percent of fisheries were considered fully exploited—harvested at or close to their maximum sustainable levels—with no room for expansion in production.[14]

Asia is the leading producer of fish, accounting for fully two thirds of global production.[15] (See Figure 3.) Latin America is a distant second, with 12 percent of world output.[16] The remaining quarter of global output is divided among the rest of the world. China is the largest national harvester of fish—its 47.5 million tons in 2008 accounted for 33.4 percent of total global production.[17] Although capture fish production in China represented just 16.5 percent of the world's total catch, aquaculture systems—including farming of salmon and shrimp—accounted for 62.3 percent of farmed fish worldwide.[18] In fact, Chinese aquaculture accounted for 23 percent of total worldwide fish production.[19]

About 81 percent of global fish production went to human consumption in 2008, and 46 percent of that came from aquaculture.[20] Overall, 27.2 million tons of fish went to non-food uses—20.67 million tons for aquaculture fishmeal alone.[21] Nearly all fish used for fishmeal come from wild capture fisheries.[22]

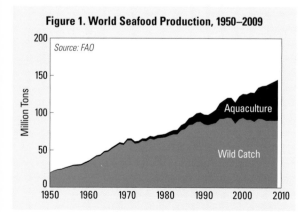

Figure 1. World Seafood Production, 1950–2009

Source: FAO

Most stocks of the top 10 marine capture species (see Figure 4)—which together account for around 30 percent of the global marine capture—were considered either exploited or over-exploited, including tuna.[23] Most stocks of anchoveta—the top capture species in 2008—were considered fully exploited, as were those of Chilean jack mackerel, Alaska pollack, Japanese anchovy, and blue whiting.

Tuna species may be the most at risk. As much as 60 percent of the world's 23 tuna stocks are fully exploited, and another 35 percent are overexploited or depleted.[24] There is substantial market demand for tuna—especially in Japan, where the popularity of tuna in sushi has skyrocketed over the past few decades.[25] Sushi's popularity has spread to the United States and Western Europe as well.[26]

The status of bluefin tuna stocks, the most valuable of all marine fish, has gained international attention in recent years because Atlantic stocks are threatened with total collapse due to overfishing; measures to regulate the catch have largely failed.[27] The Western Atlantic stock is at 18 percent of its 1970 level, and the larger Eastern Atlantic stock is at 26 percent of 1950s levels.[28] Intensive commercial fishing of bluefin tuna has continued in the Mediterranean.[29] In addition, there is growing concern that the 2010 BP oil spill in the Gulf of Mexico will have significant impact on Western Atlantic bluefin stocks.[30] Oil and the chemical dispersants that were used to break it up are known to be hazardous to fish and to their eggs and larvae.[31]

Exponential growth in the aquaculture of large predators, such as salmon and tuna, has driven overfishing of prey fish, including anchoveta and herring, which are ground into fishmeal.[32] When salmon, a carnivorous fish, are farmed, it takes at least three kilograms of fishmeal to produce one kilogram of salmon.[33] Large-scale reductions in prey species threaten the entire food chain, putting further stress on large predator stocks.[34]

The environmental status of inland fisheries is not well monitored or understood.[35] Although both commercial and subsistence fishing affect

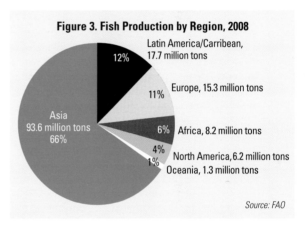

Figure 2. Global Fish Production by Sector, 2008

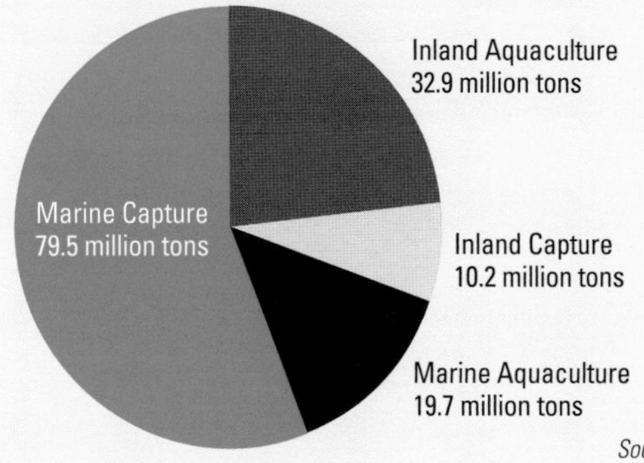

Inland Aquaculture
32.9 million tons

Marine Capture
79.5 million tons

Inland Capture
10.2 million tons

Marine Aquaculture
19.7 million tons

Source: FAO

Figure 3. Fish Production by Region, 2008

Latin America/Carribean, 17.7 million tons

12%

Europe, 15.3 million tons

11%

Asia
93.6 million tons
66%

6% Africa, 8.2 million tons

4% North America, 6.2 million tons

1% Oceania, 1.3 million tons

Source: FAO

these areas, land use practices may have an even more profound effect. Deforestation and ecosystem degradation, water use, drainage of wetlands, dam construction, and pollution can all change the composition and value of fisheries.[36]

While inland fisheries are a source of significant income and food security to many of the world's poorest people, they are consistently undervalued in national and international development agendas, according to the U.N. Food and Agriculture Organization.[37] In 2008, some 44.9 million people earned income directly from capture fisheries or aquaculture, a 167 percent increase from 1980.[38] For each direct fishing and aquaculture job there are approximately

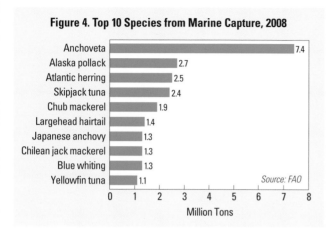

Figure 4. Top 10 Species from Marine Capture, 2008

Source: FAO

Million Tons

three secondary jobs, including post-harvest packaging and shipping to markets, adding up to 180 million fish industry jobs worldwide.[39] And each of these income earners on average supports three dependents financially. Thus the fishing industry ultimately supports the livelihoods of 8 percent of the global population.[40]

Most of the world's fishers and aquaculturists are in developing countries—with 85.5 percent in Asia alone in 2008.[41] Africa had 9.3 percent, Latin America 2.9 percent, Europe 1.4 percent, North America 0.7 percent, and Oceania just 0.1 percent.[42] There were 13.3 million Chinese fishers and fish farmers—almost one third of the world total.[43] Worldwide, employment in capture fisheries is stagnating or declining, while jobs in aquaculture are growing substantially.[44]

Nearly 60 percent of the world's fishers are small-scale commercial or subsistence fishers, also known as artisanal fishers, mainly in developing countries.[45] Hundreds of millions more engage in occasional fishing to supplement their diets and incomes.[46] Small-scale fishing can contribute significantly to poverty reduction in the developing world and is an important source of protein for many poor families.[47]

In the 1980s, poor countries—like those in sub-Saharan Africa—began selling fishing rights to wealthy European and Asian countries that had already depleted stocks off their own coasts.[48] Artisanal fishers now have to compete

with industrial fishing fleets with little or no protection by local governments.[49] In Mauritania, for example, where more than 600,000 tons of fish are caught annually, foreign fleets account for about 80 percent of the total.[50] Worldwide, wealthy countries subsidize their commercial fishers by as much as $30 billion a year, making the competition even more unbalanced.[51] Over the past 30 years, with significant pressure added by intrusive fleets, fish stocks in West African coastal fisheries have declined by up to 50 percent.[52]

It is estimated that as much as $10–23.5 billion of illegal, unreported, and unregulated fishing takes place worldwide every year.[53] In Pacific fisheries, illegal fishers harvest as much as $305 million annually in tuna alone.[54] The Secretariat of the Pacific Community estimates that Pacific fish stocks face collapse within the next quarter-century if overfishing is not adequately addressed.[55]

Top-down fisheries management has had limited success. In March 2010, the Convention on International Trade in Endangered Species of Wild Fauna and Flora (CITES) failed to pass a global ban on the trade of Atlantic bluefin tuna and several shark species.[56] And in November 2010, the countries belonging to the International Commission for the Conservation of Atlantic Tunas reduced bluefin catch quotas by only 600 million tons—well short of the 6-billion-ton reduction scientists recommend to prevent total collapse of the Eastern Atlantic bluefin stocks.[57]

Fisheries co-managed by local authorities and fishers themselves have emerged as a promising innovation.[58] More than 211 co-managed fisheries are operating worldwide.[59] In 2007, for example, a group of artisanal oyster harvesters across 15 Gambian communities formed the TRY Women's Oyster Harvesting Association.[60] By lengthening their closed season and rotating one-year off-limits zones along the Gambia River shoreline, the cooperative increased oyster numbers and sizes, securing incomes and nutrition in their communities.[61]

Marine Protected Areas (MPAs)—parts of the ocean where fishing is restricted or forbidden—have been internationally recognized as vital for sustainable fisheries management.[62] Over 5,000 MPAs existed worldwide in 2008, although much more coverage is needed.[63] MPAs allow exploited fish species to repopulate and grow to maturity.[64] Networks of national and international MPAs could work to strengthen co-managed fisheries.[65]

Over the last century, mainstream approaches to fisheries management have focused narrowly on short-term profit and increasing production. By ignoring the long-term effects of unsustainable harvesting—exacerbated by fishing subsidies—the market value of fish has remained artificially low, keeping market demand high.[66] An ecosystems approach to fisheries has emerged as a way to integrate diverse interests into fisheries management. Using this approach, the total value of a fisheries ecosystem includes the values of fish as a food product, employment in the fishing sector, sustainable mainte-nance of stocks for future use, and the ecosystems services that fisheries provide to society, such as storing and cycling nutrients and absorbing pollutants.[67] Integrating such an approach into fisheries management would give consumers a more accurate value of fish, curbing consumption.

With a growing world population and increasing demand for dietary protein, human consumption of fish is bound to continue to rise. Even as more seafood comes from fish farms, marine ecosystems are under tremendous pressure from fishing. Many fish stocks are in decline. As land-based and marine resources are stretched increasingly thin, fisheries management will need to shift to more-sustainable strategies to both meet demand and support fishers and fishing communities.

Matt Styslinger was a research intern with the Nourishing the Planet Program at Worldwatch Institute.

Meat Production and Consumption
Continue to Grow

Jesse Chang

Global meat production increased by 2.6 percent in 2010 to 290.6 million tons, an increase from the 0.8 percent growth rate of 2009.[1] (See Figure 1.) Even with this minimal increase, however, worldwide meat production has tripled since the 1970s.[2] The increase continued the steady growth of the past decade. Since 2000, global meat production has risen by 20 percent.[3]

Meat consumption is also growing worldwide. Per capita meat consumption has increased from 41.3 to 41.9 kilograms.[4] Consumption varies greatly between countries, however. In the developing world, individuals eat about 32 kilograms of meat a year.[5] But consumers in the industrial world eat about 80 kilograms per person each year.[6] (See Figure 2.)

Pork is the most widely produced meat in the world, followed by poultry, beef, and sheep.[7] (See Figure 3 and Table 1.) Total pig meat production increased by about 3 percent in 2010, to 109 million tons.[8] China, which holds nearly half of the world's pig market, has been affected by an elimination of sow subsidies—the government funds paid farmers to increase hog production—as well as by outbreaks of foot-and-mouth disease (FMD) and swine blue ear disease.[9] Reduced supplies in Asia are expected to translate into record exports by the United States to feed rising demand in traditional Asian markets such as South Korea, China, and Japan.[10]

Meat production from chickens recorded the fastest growth of all sectors, increasing by 4.7 percent to 98 million tons.[11] Global poultry exports also increased by 4 percent, with Brazil, China, the European Union, and the United States accounting for nearly two thirds of global exports.[12] China registered a 7 percent growth in poultry production in 2010.[13]

Global beef production did not grow at all, however, holding steady at 65 million tons due to low animal inventories and high prices.[14] Latin America and the Caribbean even experienced a 2 percent decrease in beef production from the previous year.[15]

Livestock play an integral role as a source of food and income for millions of people, particularly in sub-Saharan Africa. Seventy percent of the 880 million of the world's rural poor living on less than a dollar a day are partially or completely dependent on livestock for livelihoods and food security.[16] Livestock production is responsible for one third of global protein intake.[17] Demand for livestock products will nearly double in sub-Saharan Africa and South Asia, from 200 kilocalories (kcal) per person per day in 2000 to about 400 kcal in 2050.[18]

Raising livestock uses a lot of water, accounting for about 23 percent of all global water use in agriculture—this is the equivalent of 1,150 liters of water per person per day.[19] Feed crop production in developing countries requires 1–2 trillion cubic meters per year, and the produc-

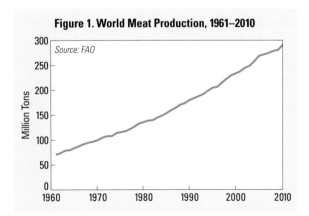

Figure 1. World Meat Production, 1961–2010

Source: FAO

tion of animals themselves takes another 536 billion cubic meters.[20]

Livestock production has also been linked to climate change. Livestock account for about 18 percent of all human-caused greenhouse gas (GHG) emissions and produce nearly 40 percent of the world's methane (a GHG 25 times more potent than carbon dioxide) and 65 percent of nitrous oxide (which is 300 times as potent as carbon dioxide).[21] A 2011 report by the Environmental Working Group found that lamb, beef, and cheese had the highest GHG emissions out of all meat products.[22] Ruminant animals generate more methane, require significantly more energy-intensive feed made from soybeans and corn, and produce more manure than other livestock.[23]

Animal diseases continue to limit total meat output.[24] Although the last few decades have seen a general reduction in the burden of livestock diseases due to better control methods, new diseases have emerged that are considerable cause for concern, particularly for human health.[25] Diseases that can infect humans include avian influenza (H5N1), swine flu (H1N1), FMD, and mad cow disease.[26] The economic impacts of livestock disease can be enormous—for example, FMD cost the United Kingdom $18–25 billion between 1999 and 2002.[27]

Many diseases are spread through industrial farming practices, which force livestock to live in crowded, dirty environments. Cramped and filthy conditions in factory farms contribute to antibiotic resistance, making it more difficult to treat human as well as animal diseases.[28] Eighty percent of all antibiotics sold in 2009 were used on livestock and poultry, meaning that just 20 percent were used for human illnesses.[29] Seventy-five percent of antibiotics are not absorbed by animals and are excreted in their waste, posing a serious risk to public health.[30] Antibiotics in animal waste that is stored or spread on fields can leech into the environment and from there into humans after they drink contaminated water.[31] Food crops are also susceptible to contamination from antibiotics found in manure.[32]

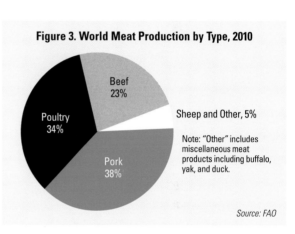

Figure 2. Meat Consumption per Person, 1961–2010

Figure 3. World Meat Production by Type, 2010

Note: "Other" includes miscellaneous meat products including buffalo, yak, and duck.

Source: FAO

Table 1. World Meat Production by Type, 2010

Meat	Production	Share of Total
	(million tons)	(percent)
Pork	109	38
Poultry	98	34
Beef	65	23
Sheep and other	13	5
Total	285	100

Note: "Other" category includes a range of smaller miscellaneous meat products, including buffalo, yak, and duck.
Source: FAO, "Meat and Meat Products," *Food Outlook, June 2009, June 2010,* and *June 2011.*

Human health can also be negatively affected by meat-reliant diets. High consumption of red and processed meats has been linked to a multitude of chronic diseases and health problems, including diabetes, obesity, cancer, and heart disease. A 2009 study of 500,000 Americans conducted by the National Cancer Institute concluded that 11 percent of deaths in men and 16 percent of deaths in women could be prevented if people ate only as much meat as the group that ate the smallest amount of it.[33]

Eating organic or pasture-raised animals can be healthier and better for the environment.[34]

Grass-fed beef typically have more nutrients and less fat, and food from free-range animals carries a lower risk of disease.[35] Organic methods of raising livestock feed and animals can also be better for the environment because they rely on less energy-intensive inputs, reduce erosion and pollution, increase carbon sequestration, and preserve biodiversity and wildlife.[36]

Jesse Chang was an intern with the Nourishing the Planet Program at Worldwatch Institute and is currently a student at Colgate University.

Global Economy and Resources Trends

Sam Beebe/Ecotrust

Rafts of timber off the coast of British Columbia, Canada

For additional global economy and resources trends, go to vitalsigns.worldwatch.org.

World's Forests Continue to Fall as Demand for Food and Land Goes Up

Bo Normander

The world's forests shrank by 1.3 percent or 520,000 square kilometers from 2000 to 2010—an area roughly the size of France.[1] (See Figure 1.) In total, forests now occupy 40.3 million square kilometers—31 percent of Earth's land surface.[2] Deforestation, mainly the conversion of forests to agricultural land, continues at a high rate in many countries. In addition, the extension of built-up areas and transport networks drives the changes in global land use.

According to the latest report from the U.N. Food and Agriculture Organization (FAO), the net loss of forests has decreased from around 83,000 square kilometers (0.20 percent) a year in the 1990s to 52,000 square kilometers (0.13 percent) a year in the first decade of this century.[3] Despite the relative slowdown in the global loss of forest, however, deforestation remains a great challenge in a number of countries and regions.

Included in the definition of forests are all types of wooded areas—tropical, temperate, and boreal, and ranging from untouched primary forests to highly productive plantations. The sta-tistics reported by FAO are based on surveys filled in by national authorities in all U.N. member countries.

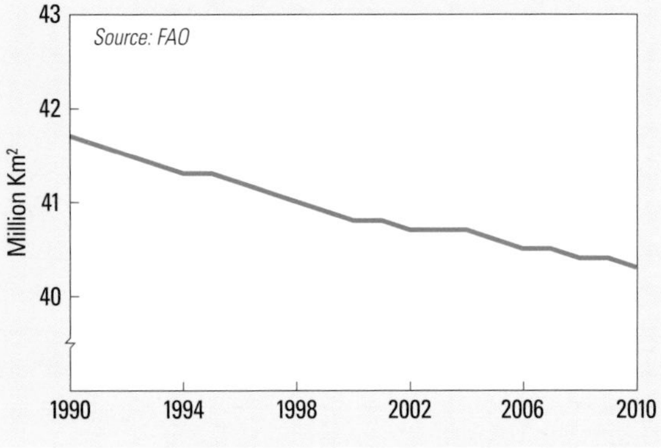

At the regional level, Africa and South America suffered the largest net losses of forests, corresponding to annual losses of 0.5 percent in both continents.[4] Africa reduced its forest area from 7.1 million square kilometers in 2000 to 6.7 million square kilometers in 2010, while South America went from 9.0 million to 8.6 million square kilometers in the same period.[5] (See Figure 2.)

The loss of forests in South America primarily reflects the trend in Brazil, which accounts for 60 percent of the forests in the region. The annual rate of deforestation in South America dropped from 44,000 square kilometers in 2000–05 to 36,000 square kilometers in 2005–10.[6] But the forest loss in the region is still alarmingly high, and Brazil tops the list of countries with the highest net loss, caused by a continual conversion of areas in the Amazon and other primary forests into plantations and farmland.[7]

In Africa, reported data indicate a persistent conversion of forest into agricultural land.[8] (See Figure 3.) Forest loss is mainly seen in the sub-Saharan region, but as few countries in Africa have reliable data from comparable assessments over time, the resulting trends should be treated with caution. Oceania also reported a net loss of forest and grasslands, mainly due to large losses in Australia, where severe drought and fires have turned forests and grasslands into barren land.[9] (In Figure 3, barren land is included in "other land" category.)

In Asia, a minor net loss of forests in the 1990s was converted into an average net gain of 22,000 square kilometers per year between 2000 and 2010.[10] This was primarily a result of large-scale afforestation reported by China, but

Figure 1. World Forested Area, 1990–2010

Source: FAO

it was also due to a reduction in the rate of deforestation in some countries, including Indonesia. That country still has the second highest net loss of forests, however, with only Brazil losing more.[11]

The historical large-scale forest loss in temperate regions has come to a halt as the forest areas in both North America and Europe have been relatively stable since the 1990s. In fact, a slight increase of 7,000 square kilometers per year between 2000 and 2010 was reported in Europe.[12] According to the European Environment Agency, the largest changes in land use in Europe are linked to the expansion of artificial surfaces, such as built-up land and transport networks.[13] Hence, the areas of agricultural land and grasslands in Europe have decreased as a result of the expansion of built-up land and afforestation.

Primary forests—in particular, tropical moist forests that show no visible signs of human intervention—include some of the world's most species-rich and diverse ecosystems. They now account for 36 percent of the world's forest area, but their area has decreased at an alarmingly high rate of 0.4 percent annually over the last 10 years.[14] This is three times faster than the annual loss for all forests. South America accounted for the largest proportion of the loss in primary forests, followed by Africa and Asia. In part, however, this is because primary forests in large parts of Europe, North America, and Asia had already been deforested or converted into productive land or plantations in historic times, and there is today a larger motivation to protect what is left of natural and primary forests in these areas.

Changes in land use and land cover are of course not a new phenomenon, but they have accelerated over the last century or two, largely driven by technological changes. It is estimated that more forests were cleared between 1950 and 1980 than in the eighteenth and nineteenth centuries combined.[15] While the forest cover decreased by at least 15 percent since 1700, cropland area almost quintupled.[16] (See Figure 4).

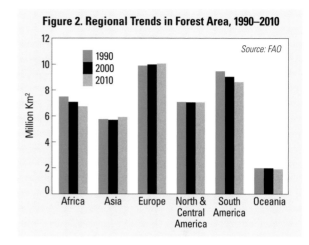

Figure 2. Regional Trends in Forest Area, 1990–2010

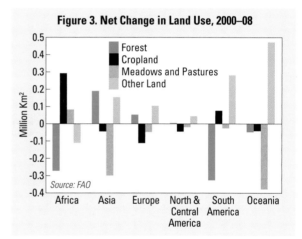

Figure 3. Net Change in Land Use, 2000–08

Replacing forest with cropland occurred at a large scale throughout the twentieth century, and projections from the Organisation for Economic Co-operation and Development (OECD) suggest that this trend will continue toward 2030.[17] (See Figure 5.) The U.N. Population Division predicts that world population will reach 9 billion by 2050, adding 2 billion people to the current population.[18] Consequently, and as a result of improved welfare in countries like China, India, and Brazil, the demand for food and fodder crops is expected to escalate, increasing the pressure to convert

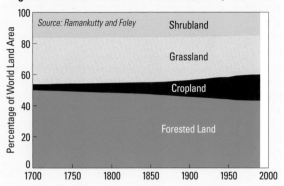

Figure 4. Historical Trends in Global Land Use, 1700–1990

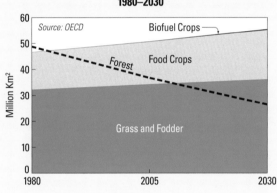

Figure 5. Projected Area of Farmland and Forest, 1980–2030

forests and other natural habitats into productive cropland.

Moreover, OECD foresees an increased demand for biofuel crops as more and more governments are determined to reduce the use of fossil fuels by replacing them with biofuels such as ethanol from maize. In the United States, 29 percent of grain production in 2010 was already used to produce fuel ethanol for the transport sector.[19]

Preserving the world's forests and natural habitats requires actions at the local, national, and global levels. At the U.N. Biodiversity Conference in Nagoya, Japan, in October 2010, governments agreed to "take effective and urgent action to halt the loss of biodiversity in order to ensure that by 2020 ecosystems are resilient and continue to provide essential services."[20] The 2020 strategic goal is accompanied by 20 headline targets. Target 5 states that "by 2020, the rate of loss of all natural habitats, including forests, is at least halved and where feasible brought close to zero." Despite this determination at the intergovernmental level to combat the loss of biodiversity and natural habitats, the driving forces behind taking more land to produce timber, food, and fodder may make it difficult to reach this target.

Bo Normander is the director of Worldwatch Institute Europe.

Tropical Forests Push Payments for Ecosystem Services onto the Global Stage

Will Bierbower

The term payments for ecosystem services (PES) describes financial arrangements and schemes designed to protect the benefits that the natural environment provides for human beings. The Millennium Ecosystem Assessment, a report of work conducted by some 1,360 scientists from around the world, estimated in 2005 that about 60 percent of all ecosystem services are being degraded or used unsustainably.[1] (See Table 1.) Governments and businesses are using PES to protect an increasing number of these services—from crop pollination to water filtration—working with a variety of stakeholders and financial arrangements. Payment schemes for watershed and biodiversity services are currently the primary markets

for ecosystem services. These markets were estimated to have a combined global value of at least $11 billion in 2008.[2] (See Figure 1.) Smaller markets exist for forest carbon sequestration programs and water quality trading.

Payments for watershed services that protect and enhance water quality were estimated to be worth at least $9.25 billion in 2008.[3] The largest market for this was in China, valued at around $7.8 billion in 2008—up from just $1 billion in 2000.[4] The second largest market was in the United States, valued at around $1.4 billion.[5]

The payments for biodiversity protection, restoration, and management were estimated to have a combined global value of $1.8–2.9 billion in 2008.[6] (By 2010, the figure reached $2.4–4.0

Table 1. Ecosystem Services: Global Status and Trends

Ecosystem Service Type	Degraded	Mixed	Enhanced
Provisioning	Capture fisheries Wild foods Wood fuel Genetic resources Biochemicals Freshwater	Timber Fiber	Crops Livestock Aquaculture
Regulating	Air quality regulation Regional and local climate regulation Erosion regulation Water purification Pest regulation Pollination Natural hazard regulation	Water regulation (e.g., flood protection) Disease regulation	Carbon sequestration
Cultural	Spiritual and religious values Aesthetic values	Recreation and ecotourism	

Source: Adapted from *Millennium Ecosystem Assessment*, Ecosystems and Human Well-being: Synthesis *(Washington, DC: Island Press, 2005)*.

Figure 1. Estimated Market Values of Top Two PES Types in 2008

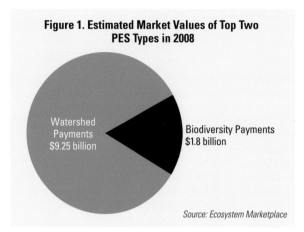

Watershed Payments $9.25 billion

Biodiversity Payments $1.8 billion

Source: Ecosystem Marketplace

billion worldwide.)[7] Most reliable data come from the United States, where payments total $1.5–$2.4 billion annually.[8] While growth has been slow in recent years in countries that already have operational programs for biodiversity protection, several other countries have been developing programs and environmental policy frameworks for biodiversity payment mechanisms. In 2010, at least 45 payment programs for biodiversity were operational across the world and 27 programs were in development, up from 39 operating programs and 25 in development in 2008.[9]

On a much smaller scale, PES carbon sequestration projects in forests were estimated to have a combined global value of around $37 million in 2008, up from $7.6 million in 2006 and down from a pre-recession high of $40.5 million in 2007.[10] The volume of transactions increased from 5.1 to 5.3 megatons of CO_2 in 2008.[11] Most of the payments for sequestering carbon in forests in 2008 involved afforestation and reforestation projects, but by 2010 projects for reducing emissions from deforestation and land degradation (known as REDD) were the majority of payments for forest carbon.[12]

PES has been formally defined as a transaction where a well-defined environmental service (or a land use likely to secure that service) is "bought" by one service buyer (or more) from one service provider (or more) on the condition

that the provider secures provision of the service.[13] The objective of PES is to encourage a net increase in benefits that would not otherwise have occurred without its financial incentive: a concept known as providing additionality.[14]

The scale of the service, the number of buyers and sellers involved, and how direct and immediate the financial payoff is can influence how a PES arrangement is constructed. PES arrangements may develop voluntarily where the ecosystem service flows directly to the buyer and there are few stakeholders involved. In the early 1990s, for instance, Perrier Vittel set up long-term contracts with farmers along a river in France that supplies the company with water to implement land use practices that reduce agricultural runoff, thus securing a cleaner water supply that directly improves the quality of its products.[15]

Often, however, the scale of the service is too large and the financial benefits too diffuse from a market perspective for users to pay for a service on their own, but protecting the service is still in the public's interest. In these cases, a government or third party can provide payments for securing the services in tandem with complementary regulations. China's Sloping Land Conversion Program is one example of such a scheme. Chinese farmers had removed vegetation in mountainous areas over several decades when converting the areas into farmland, unintentionally causing erosion of valuable topsoil and flooding downstream. In 1999, the government started paying farmers to restore the land to its original ecological state in order to reduce flooding and erosion, and now the program has become one of the largest PES schemes in the world.[16] During its first seven years, $7.7 billion was paid out to rural farmers through the program and 7.2 billion hectares of cropland were enrolled.[17]

Governments also mediate payments by setting up a banking or offset system for users to pay providers of a service. These arrangements help leverage private finance when transaction costs are too high for stakeholders to arrange deals on their own. In such cases, payments are generally transacted in exchange for output-based, measurable physical characteristics related to a service rather than input-based land use practices.[18] Wetland mitigation banks in the United States are examples of such a system. Land developers who disturb or destroy any part of a wetland are bound by law to buy credits from a bank of projects that have restored, created, or protected an area of wetland in the same watershed.[19]

Although a number of factors drive the development of PES, governments have been the key players in establishing most PES arrangements. In some areas, regulations such as those included in the U.S. Clean Water Act for wetland mitigation banking have driven demand for ecosystem services by establishing the legal obligations and procedural framework for landowners and users. As demonstrated in carbon markets, regulations developed by governments for compliance markets have the ability to drive activity up by several orders of magnitude over what voluntary markets could generate. In China, financing from the central government for the Sloping Land Conversion Program has been primarily responsible for driving the demand for ecosystem services. The design of a PES scheme is thus in part shaped by the various political, cultural, and institutional arrangements in a country or region.

In the near future, PES will likely be driven by its potential to help mitigate climate change. Its contribution would come largely through REDD—a plan to use PES to reduce deforestation, particularly in the developing world. Deforestation, which currently occurs primarily in tropical rainforest regions, accounts for an estimated 12–20 percent of human-caused greenhouse gas emissions.[20] The United Nations REDD Programme and the World Bank Forest Carbon Partnership Facility were both established in 2008 to assist the development of a global PES scheme for REDD.

An international debate is taking place over whether such a program can effectively be designed and implemented. Among the many issues under discussion are how the PES will affect the rights of indigenous peoples when it turns forests into valuable commodities, whether accounting practices encourage deforestation before a project begins, whether developing countries have the capacity to monitor forest carbon stocks and properly enforce forest regulations, whether there is the technical capacity to prove that projects are providing additional benefits that would not otherwise have occurred, whether participants in REDD programs can ensure that forests will remain standing in perpetuity, and whether protecting forests in one area would drive deforestation into another area.[21]

The international community has discussed scaling up REDD finance to $30 billion per year, but funding remains tight as countries such as the United States have voted down legislation for national carbon markets that could potentially finance the program.[22] California is set to initiate in 2013 the first compliance cap-and-trade market that will accept certified emissions reductions from REDD projects under its provision for offsets, creating the possibility that regulated emitters in California working to meet emissions reduction goals will fund the projects.[23] Meanwhile, several governments, including Norway and Germany, continue to provide financial support for the Forest Carbon Partnership Facility and REDD to help countries develop the technical and governance capacity to handle a REDD market as well as to provide financial incentives to the best performers.[24]

Will Bierbower was a Climate and Energy research intern at Worldwatch Institute. He wishes to thank Sam Shrank and Daniel Kandy for their help with this article.

Value of Fossil Fuel Subsidies Declines, National Bans Emerging

Alexander Ochs and Annette Knödler

Global fossil fuel consumption subsidies fell to $312 billion in 2009 from $558 billion in 2008, a decline of 44.1 percent.[1] The reduction is due primarily to changes in international energy prices as well as in domestic pricing policies and demand, rather than because the subsidies themselves were curtailed. The number also does not include fossil fuel production subsidies that aim at fostering domestic supply, which are estimated at an additional $100 billion globally per year.[2]

Fossil fuel consumption subsidies include public aid that directly or indirectly lowers the price for consumers below market price. The International Energy Agency (IEA) defines energy subsidies as "any government action directed primarily at the energy sector that lowers the cost of energy production, raises the price received by energy producers or lowers the price paid by energy consumers."[3] Common means of subsidizing energy include trade instruments, regulations, tax breaks, credits, direct financial transfers like grants to producers or consumers, and energy-related services provided by the government, such as investments in energy infrastructure or public research.[4] Many observers believe that fossil fuel subsidies should be phased out because they reduce the competitiveness and use of cleaner, alternative energy sources.

The energy sources that are most heavily subsidized are unquestionably oil products ($312 billion in 2008) and natural gas ($204 billion in 2008).[5] Coal received $40 billion.[6] (See Table 1 and Figure 1 for profiles of fossil fuel subsidies by country.) In countries that belong to the Organisation for Economic Co-operation and Development (OECD), production subsidies are by far the most prevalent form of subsidization. In other countries, subsidies are more likely to go to consumers. A phaseout of subsidies of either kind will most strongly affect producing countries in the Middle East, Russia, and parts of Asia.[7]

Global fossil fuel subsidies are roughly an order of magnitude greater than subsidies to other energy technologies. Subsidies for nuclear power, for example, totaled $45 billion in 2007.[8] Support for renewable energy and biofuels totaled $47 billion in 2007.[9]

International institutions, not just governments, are also involved in providing subsidies. In November 2010, the World Bank reported a record high for fossil fuel funding that year of about $6.6 billion, almost twice as much as for renewables and efficiency ($3.4 billion).[10] Coal funding accounted for 66 percent of the Bank's overall energy funding in 2010.[11] Between 2005–07 and 2008–10, World Bank support for fossil fuels tripled, while spending on renew-

Table 1. Fossil Fuel Consumption Subsidy Rates as Proportion of Full Cost of Supply, Top 10 Countries, 2009

Country	Average Subsidization
	(percent)
Iran	89.2
Kuwait	83.3
Saudi Arabia	78.9
Venezuela	72.0
Turkmenistan	66.9
Qatar	63.2
Uzbekistan	56.7
Egypt	56.3
United Arab Emirates	55.7
Libya	52.0

Source: IEA, World Energy Outlook 2010, at www.iea.org/ subsidy/index.html, 2010.

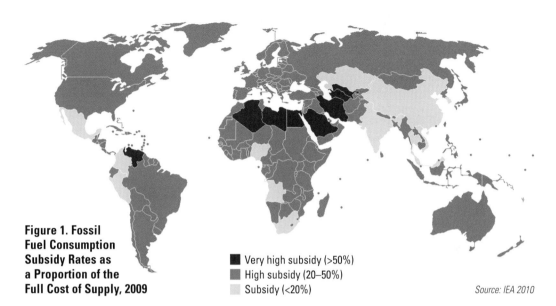

Figure 1. Fossil Fuel Consumption Subsidy Rates as a Proportion of the Full Cost of Supply, 2009

■ Very high subsidy (>50%)
■ High subsidy (20–50%)
□ Subsidy (<20%)

Source: IEA 2010

ables increased fivefold.[12] On a per-kilowatt-hour basis, the numbers change significantly. In 2007 renewables were subsidized at 5¢/kilowatt-hour (kWh), compared with nuclear energy, which received 1.7¢/kWh, and fossil fuels at 0.8¢/kWh.[13]

But not only do subsidies tend to be higher for newer energy technologies than for established ones (the global share of renewables in final energy consumption in 2008 was 19 percent), shifting official support from fossil fuels to renewables is thought to be essential for decarbonizing the global energy system.[14] Such a shift could help create a triple win for national economies. First, GHG emissions would decline. The IEA shows that a cut in fossil fuel subsidies will reduce carbon dioxide (CO_2) emissions by 5.8 percent, or 2 gigatons, in 2020 compared with a baseline case in which subsidy rates remain unchanged.[15] Second, long-term economic growth would be spurred, creating new jobs. In Germany alone, some 340,000 green jobs have been created over the last few years.[16] And third, national security would be strengthened as dependence on imported energy lessened. In Germany, as of 2020, more than 20 billion euros will stay in the country annually

due to reduced energy imports.[17] This number is expected to increase to 40 billion euros in 2030.[18]

The Berlin-based group Green Budget Europe estimates that Germany could save 6.2 billion euros annually when cutting indirect subsidies through linking tax deductibility of company cars to their CO_2 footprint, abolishing privileges in the air traffic through introducing a ticket fee, and raising the heavy-vehicle toll.[19] *Green Scissors 2010*, a report by Friends of the Earth and other U.S. groups, pursues the same approach and suggests a $109 billion cut in environmentally harmful government spending in the agricultural, infrastructural, and public lands sector.[20] The external costs of energy production could also be added as indirect subsidies for fossil fuels. (See Figure 2.) The costs of cleaning up air and water pollution and of pollution-related health care in China, for example, amount to 5.8 percent of the country's gross domestic product (GDP).[21] The widely cited 2006 Stern Review, *The Economics of Climate Change*, shows that the costs of non-action would rise by 5–20 percent of gross world product annually.[22]

The impact of phasing out fossil fuel subsidies would be significant. IEA estimates that a

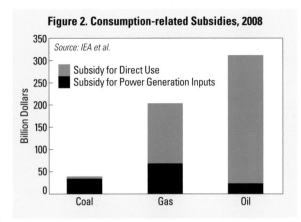

Figure 2. Consumption-related Subsidies, 2008

a program of the International Institute for Sustainable Development, is convinced that the removal of fossil fuel subsidies would lead to a real increase in GDP everywhere, ranging between 0.1 percent in total by 2010 and 0.7 percent a year to 2050.[25] However, a study by IEA, OPEC, OECD, and the World Bank shows that the real income gains of a multilateral removal of fossil fuel subsidies would be distributed unequally to poorer producers, especially fossil-fuel-producing countries like Russia and East European countries that are not in the European Union (EU).[26] (See Figure 3.)

At the Seoul Summit of G20 nations in 2010, the heads of government renewed their pledge to phase out inefficient fossil fuel subsidies, echoing their 2009 statement at the Pittsburgh Summit: "Inefficient fossil fuel subsidies encourage wasteful consumption, reduce our energy security, impede investment in clean energy sources and undermine efforts to deal with the threat of climate change."[27]

global removal by 2020 would lead to a cut in global primary energy demand of 5 percent (compared with a baseline case in which subsidy rates remain unchanged), which is equivalent to the current consumption of Japan, South Korea, and New Zealand.[23] Further, IEA estimates show that oil demand would be reduced by 4.7 million barrels a day by 2020—6 percent of 2008 consumption—compared with a baseline case in which subsidy rates remain unchanged.[24] The Global Subsidies Initiative,

Several governments already have plans to phase out fossil fuel subsidies:

• The European Commission has adopted a 2020 strategy that reemphasizes earlier calls on member states to phase out environmen-

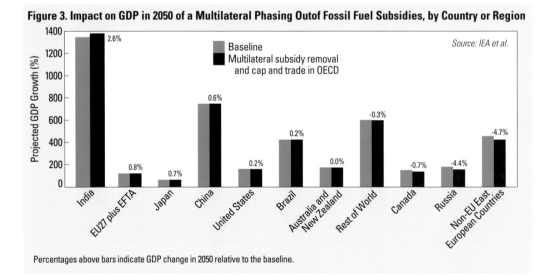

Figure 3. Impact on GDP in 2050 of a Multilateral Phasing Outof Fossil Fuel Subsidies, by Country or Region

Percentages above bars indicate GDP change in 2050 relative to the baseline.

tally harmful subsidies.[28] The EU is also committed to fully phasing out coal subsidies by 2018.[29]

- In its 2011 budget, the U.S. government proposed elimination of funding for programs that provide "inefficient fossil fuel subsidies that impede investment in clean energy sources and undermine efforts to deal with the threat of climate change."[30]
- Indonesia started its fuel subsidy reform in 2000, which was followed by fossil fuel prices increases between 2000 and 2008.[31] In order to back up the reforms, the government provided cash transfers to the poorest (the Bantuan Langsung Tunai Program). In April 2010, the Indonesian Ministry of Energy and Mineral Resources announced it would cut fuel subsidies by 40 percent by 2014.[32]
- The Indian government announced a stop in gasoline subsidies in June 2010; the more heavily subsidized diesel, natural gas, and kerosene will continue with only a moderate reduction.[33] Nevertheless, India committed to phasing out diesel subsidies over time.[34]
- Iran reduced its energy subsidies dramatically in December 2010, at the start of a five-year

program to bring prices in line with international market levels.[35]

- The United Arab Emirates reduced gasoline subsidies in 2010; diesel prices are already largely deregulated.[36]
- Electricity subsidies are due to be phased out in Pakistan; tariffs increased by about 20 percent in 2010.[37]
- In January 2011, Syria announced it would expand its subsidy program by increasing heating oil allowances.[38]
- Malaysia started a reform program in 2010, reducing subsidies for petrol, diesel, and liquefied petroleum gas.[39]
- South Africa aims to increase electricity tariffs by approximately 25 percent a year between 2010 and 2013.[40]

Although these are all promising trends, the world is still far from being free of fossil fuel subsidies.

Alexander Ochs is director of the Energy and Climate Program at the Worldwatch Institute. Annette Knödler, a graduate student at the University Viadrina in Frankfurt/Oder, Germany, was an intern in the program.

Energy Intensity Is Rising Slightly

Haibing Ma

Global energy intensity, defined as worldwide total energy consumption divided by gross world product, increased 1.35 percent in 2010.[1] (See Figure 1.) Since the global financial crisis in 2008, worldwide energy consumption has grown faster than the global economy for two years in a row, as many countries started implementing massive stimulus packages to push

their national economies out of recession.[2]

This rising energy intensity reverses the broader trend of the last three decades. From 1981 to 2010, global energy intensity decreased by about 20.46 percent—about 0.79 percent each year.[3] During this period, most industrial countries restructured their economies, with energy-intensive heavy industry accounting for a progressively smaller share. New technologies applied to energy production and consumption significantly improved efficiency in almost every sector of the economy. The past few decades have also seen some energy-intensive industries migrate from industrial countries to emerging economies like China and India, but such transfers were largely based on mature technologies and already improved efficiency, which helped maintain the declining trend of energy intensity at the global level.

Yet the pace of global energy intensity improvement has slowed in the last 10 years. Energy intensity dropped at an average annual rate of 0.98 in the 1980s and at 1.40 percent in the 1990s.[4] During 1991–2000, the surge of the so-called knowledge-based economy significantly boosted global economic productivity without consuming too much energy.[5] Then from 2001 to 2010, energy intensity dropped more slowly, at 0.03 percent a year on average.[6]

During this last decade, energy intensity has actually fluctuated. A slight rise between 2002 and 2004 was followed by a drop of 0.8 percent, much faster than the average for the entire decade.[7] (See Figure 2.) This fluctuation looks unusual, given that the previous two decades never showed this kind of trend. But if real gross domestic product (GDP) values based on purchasing power parity (PPP) are used, the fluctuating "S" shape disappears, and global energy intensity in the last 10 years still follows the

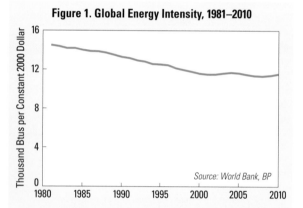

Figure 1. Global Energy Intensity, 1981–2010

Source: World Bank, BP

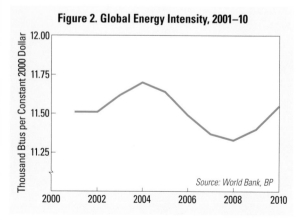

Figure 2. Global Energy Intensity, 2001–10

Source: World Bank, BP

overall declining trend.[8] (See Figure 3.)

Using GDP values expressed in a PPP-based conversion ratio, global energy intensity exhibits three trends: energy efficiency worldwide was constantly increasing until recently; between 2004 and 2008, with an average annual rate of 1.87 percent, global energy intensity experienced its sharpest decline in 30 years; and starting in 2009, worldwide energy intensity rose for the first time in three decades.[9] In addition to technological advances, the rise in energy prices contributed to the sharp decline in global energy intensity between 2004 and 2008. Worldwide crude oil prices more than quadrupled, the fastest rise since the oil crisis in the late 1970s, which played a key role in dampening worldwide energy demand.[10] But after international oil prices dropped 75 percent in the second half of 2008, global energy intensity started rising.[11]

Advanced economies like the United States, Germany, and Japan followed the global trend of declining energy intensity. Compared with the relatively sharp decline in the United States and Germany, Japan made more modest progress.[12] (See Figure 4.) The country even experienced an increase in energy intensity during the 1990s, which may largely be attributed to the economic recession there. Japan has long been regarded as one of the most energy-efficient countries in the world.[13] But when fluctuations in currency exchange rates are eliminated by using PPP value-based GDP data, Germany surpassed Japan in the early 1990s and has since maintained its leadership in energy efficiency.[14]

China may have made the most progress in this field, with a 65 percent decline in energy intensity in the past 30 years.[15] Between 1981 and 2002, China's energy intensity dropped by 4.52 percent annually.[16] Following a brief rise after 2002, when energy-intensive industries expanded rapidly, China's energy intensity dropped 15.37 percent during 2005–10, although that fell short of the government's goal of 20 percent.[17] One reason for the shortfall was that more than half of the 4 trillion RMB ($630 billion) stimulus plan was invested in infra-

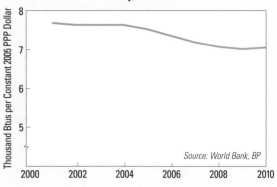

Figure 3. Global Energy Intensity, at Purchasing Power Parity, 2001–10

Source: World Bank, BP

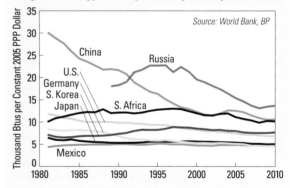

Figure 4. Energy Intensity Trends by Country, 1981–2010

Source: World Bank, BP

structural development, which drove up overall energy consumption.[18]

The most turbulent energy intensity trends were witnessed in newly industrialized and transitional countries. South Korea, for instance, had an increase in energy intensity as its industry-heavy national economy achieved rapid growth in most of the 1980s and 1990s. The country's industries were hit hard by the 1997 Asian financial crisis, which caused a sharp reversal of the growing trend of energy intensity.[19] Since the early 2000s, however, Korean government and industry started shifting the growth pattern by focusing more on advanced-technology-related research and development and on environment-friendly green energy initiatives. These efforts not only helped the country regain its economic vitality, they also contributed to lower energy intensity. Similarly, Russia experienced a rise and then a decline of energy intensity as it slowly recovered from economic recession in the 1990s.[20]

Global energy intensity is likely to keep rising in the next couple of years as the world continues to rely on large-scale infrastructure development as a way to create more jobs and bring the global economy out of recession. In the long term, as a new international climate framework is being shaped, more countries will have a

stronger incentive to transition toward more climate- and environment-friendly development patterns. Such a green transition could boost new industries such as clean tech and renewable energy as the new economic growth engines. If that happens, not only would global energy intensity continue its declining trend, but the world could achieve more sustainable development when cleaner energy sources account for a larger share of total energy consumption.

Haibing Ma is the manager of the China Program at Worldwatch Institute.

Population and Society Trends

Women in a Chinese factory assemble and test fiber optic systems

For additional population and society trends, go to vitalsigns.worldwatch.org.

Steve Jurvetson

World Labor Force Growing at Divergent Rates

Elizabeth Leahy Madsen

The world's potential labor force—measured as men and women between the ages of 15 and 64—stands at 4.6 billion people in 2011, up 17 percent over the last decade.[1] The potential labor force has tripled since 1950, and people of working age now account for nearly 66 percent of the total population—the highest ratio since 1950.[2] (See Figure 1.) This ongoing growth in the potential labor force has both positive and negative implications: there are more potential workers to drive economic expansion, but the number of available jobs may not keep pace. Given the current economic downturn, the International Labour Organization (ILO) estimates that the ranks of the unemployed reached 205 million people in 2010—a global unemployment rate of roughly 6 percent.[3]

While the overall picture is one of continued but slowing growth in the potential labor force, and hence a growing need to maintain employment rates worldwide, trends are heading in quite disparate directions in different regions and countries. In developing countries, where women have on average nearly three children, the potential labor force has grown much more rapidly since 1950.[4] Growth of 270 percent in these regions has far surpassed the rate of nearly 60 percent in industrial countries, where fertility rates on average are already lower than the replacement level needed to sustain a population at a steady level.[5] This is reflected in the disparities between the proportional size of regional labor forces and economies. The high-income countries are home to 16 percent of the world's potential labor force, yet they produce more than two thirds of the world's gross domestic product.[6] Meanwhile, nearly one quarter of the world's potential labor force lives in South Asia, but the region's share of the global economy is just over 3 percent.[7] (See Figure 2.)

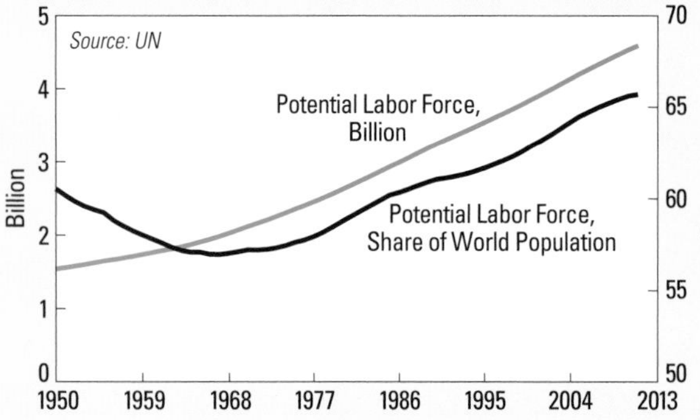

The labor force is called "potential" because not everyone between the ages of 15 to 64 holds a job or earns income, for reasons such as schooling, child or elder care, unemployment, social custom, early retirement, or poor health or disability. The ILO calculates the difference between the size of the potential labor force and the economically active population, which is also known as the labor force participation rate.[8] For the world as a whole, there were 3.2 billion economically active people age 15 to 64 in 2011, or about 70 percent of the potential labor force.[9] Fewer women than men are active in the labor force; this "gender gap" is largest in the Middle East and North Africa and in South Asia, and it is smallest in high-income countries and in East Asia and the Pacific.[10]

Countries with a high share of their population in the potential labor force have gone through a pronounced demographic shift: fertility rates have declined fairly recently, the proportion of older adults remains low, and women with smaller families are more likely to enter

Figure 1. Size of World Potential Labor Force, 1950–2011

Source: UN

Potential Labor Force, Billion

Potential Labor Force, Share of World Population

the workforce. During this period of demographic change, countries can experience significant economic benefits if their growing labor forces are healthy and educated and if government policies promote savings, investment, and open economies. Research has shown that the growth in the share of the population in the potential workforce accounted for 25–40 percent of the economic boom experienced by the rapidly growing East Asian countries between 1965 and 1990.[11] An educated, well-trained workforce and government policies that promote economic diversification and financial investment are particularly important for achieving this possible "demographic dividend" because countries with a higher share of educated and trained workers have higher productivity and economic growth rates.[12]

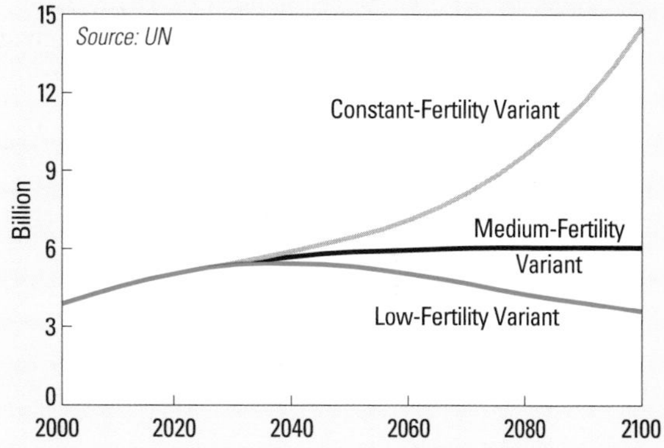

The world's potential labor force is projected to continue growing, although at a slower pace, through the second half of the century.[13] (See Figure 3.) If the total fertility rate across the world falls from 2.5 to just above 2.1 children per woman by 2050, as in the UN's "medium" variant projection, the potential labor force is expected to reach 5.9 billion people by mid-century—an increase of 28 percent relative to 2011.[14] Even if fertility rates were to fall much lower, the power of demographic momentum would still drive the potential labor force to 5.3 billion, a 15 percent increase.[15] By 2100, varying possible fertility paths create a wide range of possibilities in the size of the potential labor force. If the average global fertility rate falls gradually to 2.0 by 2100, momentum would still push the potential labor force to plateau at 6 billion.[16] However, if fertility falls to the very low level of just over 1.5 children per woman by 2100, the potential labor force would peak in the 2030s and drop to 3.6 billion by the turn of the century.[17] In the unlikely event that fertility rates remain unchanged from today's levels, the potential labor force would more than triple, to 14.5 billion, by 2100.[18]

Apart from absolute numbers, the potential

Figure 2. Shares of World Labor Force and GDP, by Region, 2010

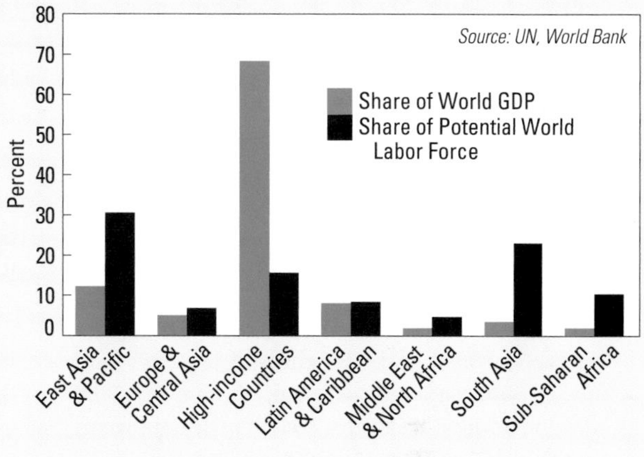

Figure 3. Projected Potential World Labor Force, 2000–2100

labor force's share of total population is projected to peak within the next five years. By mid-century, the potential labor force is projected to gradually fall to 63 percent of the total population, the same share as seen in 2000–01.[19] As fertility rates decline, however, older workers will represent a larger portion of the potential labor force, whereas in recent years the potential workforce has been more concentrated among young adults.

Labor force differences due to regionally varying fertility rates are likely to become more pronounced, as by 2050 industrial countries will see a decline in the size of their potential labor forces for the first time.[20] Meanwhile, the potential labor force in developing countries is projected to grow by an average of 39 percent between 2010 and 2050.[21] (See Figure 4.) This growth will be concentrated in the areas with the highest fertility rates—sub-Saharan Africa, the Middle East, and South Asia.[22] Although many of these countries' economies are expanding rapidly, their governments and industries face a major challenge in keeping job growth on pace with demographic trends. In Uganda, which has one of the highest fertility rates (over six children per woman) and where nearly half of the population is younger than 15, approximately 100,000 new jobs were generated in 2009.[23] But if the fertility rate remains stable at a high level, as it has for decades, the country would need to generate over 1.5 million new jobs annually by the late 2030s.[24]

Declines in the size of the potential labor force as the effects of below-replacement fertility take hold are projected for most of Europe (although not in northern Europe, due to its higher fertility rates) and the more industrialized countries in East Asia.[25] Two of the world's largest economies—Germany and Japan—are already experiencing declining workforces; in

Japan, the potential labor force could shrink by more than one third by 2050 if the fertility rate stays near 1.3 children per woman.[26] In the United States, if fertility rates stay at replacement level, the potential labor force is projected to grow slightly.[27]

The two largest countries in the world, China and India, are projected to experience quite different tracks in the demographics of their labor markets. (See Figure 5.) China's potential labor force has doubled in size since the early 1970s and stands just shy of 1 billion people, representing 72 percent of its total population.[28] However, the rapid fall in fertility from nearly three children per woman in 1980 to 1.6 today is projected to soon cause a peak and then a decline in the size of the potential labor force.[29] If fertility continues to decline slightly over the next 10 years, the potential labor force will begin shrinking by 2020.[30] Even if fertility rebounds slightly, as the United Nations projects, China's economy will be powered by a potential labor force that is nearly 20 percent smaller by 2050.[31]

In India, the fertility rate has fallen more recently and gradually, and it still remains high enough, at more than 2.5 children per woman, to generate continued population growth for decades.[32] The current potential labor force of 800 million represents 64 percent of the country's still-youthful population.[33] But even though fertility rates are likely to continue falling, the potential labor force is projected to grow, albeit at a slower pace, into mid-century.[34] By the 2030s, the potential labor force could grow by 300 million people, an increase that could serve as an engine for economic growth and development if it is matched with growth in education and employment opportunities.[35]

Labor force participation rates—the share of the potential labor force that is economically active—also vary dramatically between the two countries. In China, 74 percent of all adults over age 15 are economically active, well above the global

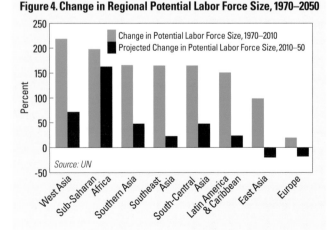

Figure 4. Change in Regional Potential Labor Force Size, 1970–2050

Legend:
- Change in Potential Labor Force Size, 1970–2010
- Projected Change in Potential Labor Force Size, 2010–50

Source: UN

(Regions: West Asia, Sub-Saharan Africa, Southern Asia, Southeast Asia, South-Central Asia, Latin America & Caribbean, East Asia, Europe)

average of 65 percent; in India, labor force participation is much lower at 58 percent.[36] This gap is in large part due to the fact that just one third of women in India work outside the home, compared with two thirds in China.[37] In addition, although their economies are growing and diversifying rapidly, both countries' labor markets still rely on unskilled, low-paying jobs.

Changes in the size of a country's potential labor force are affected not only by fertility rates but also by the level of international migration. Historically, rapidly growing working-age populations in developing countries with poor job prospects have motivated people to migrate in search of work. International migrants account for 10 percent of the total population of industrial countries, which have seen the number of migrants increase 55 percent since 1990.[38]

Although it cannot completely offset the effect of sustained declines in fertility rates, migration has a mitigating effect on population aging in industrial countries. The scale of this effect depends on governments' openness to receiving immigrants. For example, Germany is home to the third-highest number of international migrants in the world, following the United States and Russia.[39] Between 2000 and 2010, the country gained nearly a million immigrants, equivalent to 1 percent of the total population.[40] Japan, which has more restrictive immigration policies, saw its migrant population increase by half a million in the same time period, or 0.4 percent of the population.[41] If both countries' fertility and migration rates hold steady, the share of Japan's population over age 65 would be 5 percentage points higher than Germany's by 2050.[42]

Although many policymakers have expressed alarm about health care and pension system costs from an aging workforce in industrial countries, there is significant variation in the age at which people stop working. Already, more than one fifth of people age 65 or older are economically active, especially in developing coun-

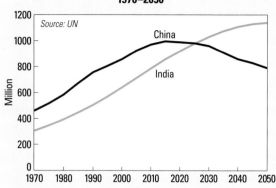

Figure 5. Potential Labor Force Size of China and India, 1970–2050

tries.[43] More than half the men over age 65 in Africa are still in the labor force.[44] In industrial countries, where life expectancies are longer, many people are also working past traditional retirement age; in Japan and South Korea, men work on average to age 70.[45] In Germany, Italy, and Spain, however, where populations are aging quickly, men tend to stop working in their early 60s.[46] Policies that promote an extension of working years for healthy and productive older adults may help offset the economic consequences of demographic change.[47]

The demographics of global and regional potential labor forces, which can be relatively confidently predicted for the near term, will have important implications for the world economy. Many developing countries will face the challenge of expanding their labor markets to provide jobs for a growing workforce. Meanwhile, industrial countries will face important policy decisions about productivity in an aging workforce and about their openness to immigration.

Elizabeth Leahy Madsen is an independent consultant and researcher specializing in demographic trends and their policy implications. She would like to thank Carl Haub for reviewing an earlier draft of this piece.

Women Slowly Close Gender Gap with Men

Robert Engelman

Women's well-being and social participation relative to men improved steadily though slowly and unevenly during the last half of the past decade, according to the "gender gap" index developed by the World Economic Forum. The index's median country score in 2010 was 68 percent, meaning that in half of the 134 countries in the index, women had closed more than that percentage of the possible gender gap with men in health, education, and economic participation and opportunity.[1] In the other half they had closed less than 68 percent of the gap. But gender gaps vary widely around the world, and in a few countries they have worsened in recent years.

Since 2006, the World Economic Forum has been tracking gender gaps in four broad categories—health, education, economic participation, and political participation—in dozens of countries for which data can be differentiated by sex. Each autumn, the Forum releases the country rankings based on an index with scores on how closely women approach men in each of these categories. The Forum report does not provide a single aggregate gender-gap score for the world as a whole, but it does average gender gaps in each of the four categories. And it counts the countries in which gaps have narrowed and those in which they have widened.

In 2010, Iceland was the best country for women's equality, the index found, with a score of 85 percent. This indicates that women experience social outcomes that average 85 percent as good as the outcomes for men across all categories and indicators within each category. Other Nordic countries—Norway, Finland, and Sweden—were right behind Iceland. In the bottom-ranked country, Yemen, with a score of 46 percent, women's social outcomes average less than half those of men.

Regional differences in the overall gap between male and female social outcomes are significant. North America (the United States and Canada) had the narrowest average gender gap among six regions, at about 74 percent. Europe and Central Asia followed, with 71 percent. Next was Latin America and the Caribbean, with 67 percent; Asia and the Pacific, with 64 percent; Sub-Saharan Africa, with 63 percent; and finally the Middle East and North Africa, with 59 percent.

All but 16 countries among 114 tracked since 2006 have seen a narrowing of the gaps between male and female outcomes during the period, according to the *Global Gender Gap Report 2010*. Gaps in female and male outcomes widened, generally modestly and with some reversals, in such low-income countries as Tanzania and high-income ones such as Sweden. The aggregate gender gap in the United States rose slightly overall from 2006 to 2010, though it declined in 2007 and 2009.

The 2010 report also included aggregate gender gap scores from 2000 for 39 countries for which the needed data were available. In all 39, gender gaps narrowed over the decade, although such countries as Colombia, the United Kingdom, and Italy experienced significant declines in some years. (See Figure 1.)

In general, countries with higher levels of income have narrower gender gaps, while poorer countries have wider ones. Still, there are some surprises, such as Lesotho in southern Africa and the Philippines, ranked eighth and ninth worldwide. The U.S. score of 74.1 percent placed it in nineteenth place among countries, behind both Lesotho and the Philippines as well as South Africa (75.4 percent) and Sri Lanka (74.6 percent). Mozambique, although it has among the lowest per capita incomes in the world, ranked twenty-second, with a score of 73.3 percent—not far behind the United States. Most country gender gaps have improved at roughly similar rates regardless of what their level was when the index first tracked them.

Worldwide, women are approaching equality with men in health and education. In the 134 countries covered in the 2010 index, women experience positive health outcomes that are 96 percent of those of men when populations are weighted by size and the indicators are adjusted for natural differences among the sexes. However, the sole health indicators the Index uses are natural birth ratios and expected years of healthy life. Only China and India and handful of other countries are known to have birth ratios significantly skewed by sex-selected abortion.[2]

Among the nearly five dozen countries not yet included in the index are Sudan, Bosnia, Serbia, Afghanistan, Somalia, Myanmar, Haiti, Bhutan, and such island states as Papua New Guinea, Vanuatu, and Micronesia. Many of these countries have small populations and poorly functioning governments that produce little publicly available data on their populations. Nonetheless, it is quite possible that if data on the 10 percent of the world's population not included in the *Gender Gap Index* existed and were included, the global outcome would be different. And the same can be said about having more indicators included. Some 350,000 women die each year, for example, from causes related to pregnancy and childbirth.[3] That number would need to be reduced to zero to close the health gap with men if this indicator were included in the index.

Women's educational attainment and literacy rates average 93 percent that of men's worldwide, according to the report. The gender gaps remain wide, however, in economics: women globally have achieved

Figure 1. Gender Gap, 39 Countries, 2000, 2005, and 2010

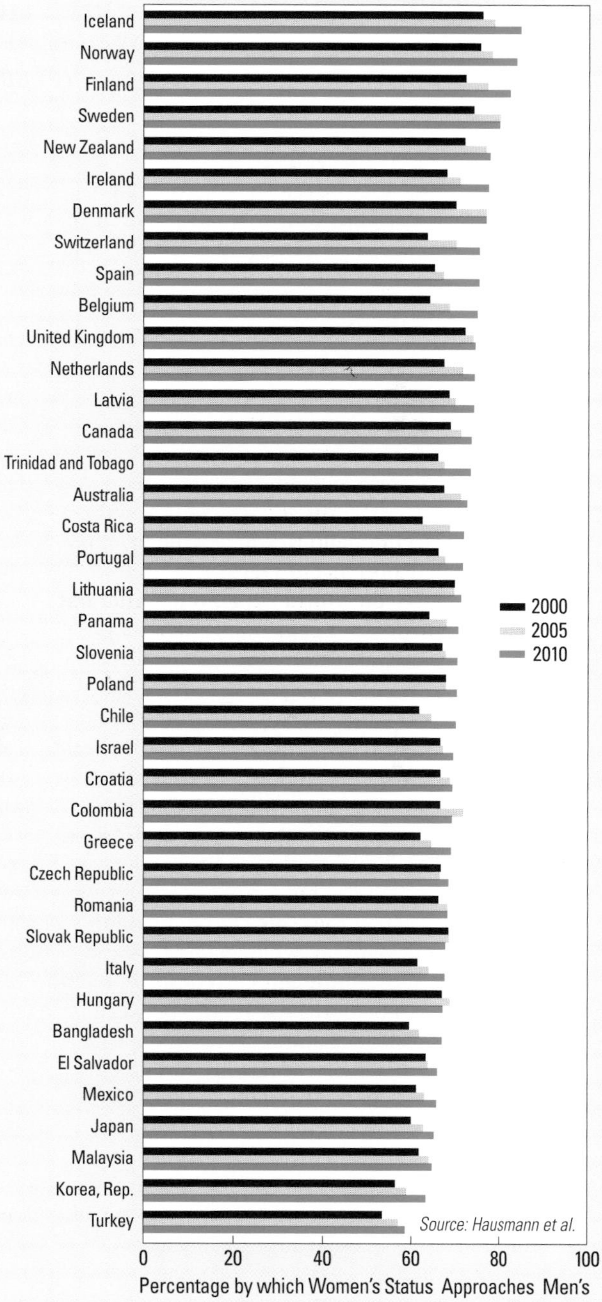

Legend: 2000, 2005, 2010

Countries (top to bottom): Iceland, Norway, Finland, Sweden, New Zealand, Ireland, Denmark, Switzerland, Spain, Belgium, United Kingdom, Netherlands, Latvia, Canada, Trinidad and Tobago, Australia, Costa Rica, Portugal, Lithuania, Panama, Slovenia, Poland, Chile, Israel, Croatia, Colombia, Greece, Czech Republic, Romania, Slovak Republic, Italy, Hungary, Bangladesh, El Salvador, Mexico, Japan, Malaysia, Korea, Rep., Turkey

x-axis: 0, 20, 40, 60, 80, 100

Percentage by which Women's Status Approaches Men's

Source: Hausmann et al.

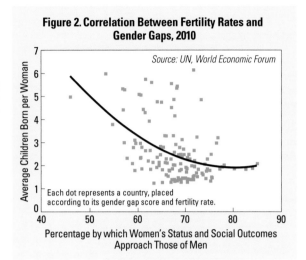

Figure 2. Correlation Between Fertility Rates and Gender Gaps, 2010

Source: UN, World Economic Forum

Average Children Born per Woman

Each dot represents a country, placed according to its gender gap score and fertility rate.

Percentage by which Women's Status and Social Outcomes Approach Those of Men

just 59 percent of equality with men. And the gap is abysmal in politics, where women's representation is 18 percent of men's.

Women's health, educational, economic, and political positions relative to men's have a great deal to do with fairness to women but relate to overall human well-being as well. The *Gender Gap Index* is highly correlated with an index on global economic competitiveness that the World Economic Forum also publishes.[4] Similar correlations were found between high scores in the gender gap index and high gross domestic product per capita, based on data published by the World Bank.[5]

One correlation the report did not explore was the connection between gender gaps and fertility, the most important current force in differences in population growth rates between countries. That correlation, however, appears to be strong, based on the fertility estimates and projections of the Population Division of

the U.N. Department of Economic and Social Affairs.[6] (See Figure 2.) Note that in most countries the precise 2010 fertility rates are not known with certainty until well after the year ends. The country fertility data used here are extrapolated from U.N. estimates and projections covering the years 2005 through 2015. In general, and with some variance (indicated in cases where dots representing countries are well above or below the regression trend line), the narrower the gender gap, the lower the fertility. This suggests an important connection to rates of population growth and, ultimately, to the environmental implications.

Some of the lowest fertility rates in the world are in industrial countries where the gender gap index score is unimpressive. A notable case is Japan, whose fertility rate is an average of 1.27 children born in a woman's lifetime—well below the 2.1 children that over time sustains a stable population size—but which ranks ninety-fourth in the gender gap index, with a score of 65.2 percent. Such hyper-low fertility is sometimes attributed to women's lack of support from partners, employers, and society at large in raising children. The index suggests that it is possible that in countries with the narrowest gender gaps in the world, fertility begins rising toward the "replacement rate" of an average of slightly more than two children per woman. The reasons for this are not clear but presumably reflect outcomes and status for women that encourage this level of childbearing. Indeed, the four Nordic countries in the index with the highest scores had progressively higher fertility rates as their scores advanced, with Iceland at 2.08 children per woman, essentially replacement fertility.

Robert Engelman is president of Worldwatch Institute.

Numbers of Overweight on the Rise

Richard H. Weil

The number of adults worldwide who are overweight jumped from 1.454 billion in 2002 to 1.934 billion in 2010, an increase of 25 percent.[1] (See Table 1). Some 23 percent of individuals age 15 or older were overweight in 2002, while in 2010 the figure rose to 38 percent—even though the number of adults increased by only 11 percent during these eight years.[2] Much of this change occurred in the industrial world. Economic, cultural, and possibly genetic factors all played a part. But in every country where people have gotten heavier the result has been the same: an increase in preventable medical problems.[3]

"Overweight" is used here for people with a body mass index (BMI)—a measure relating a person's height to weight—of 25 or greater.[4] (A person with a BMI of 30 or above is usually labeled "obese," but here the term overweight covers overweight and obese populations combined.)[5] The data considered are for those 15 and older in 177 nations—home to the vast majority of the world.[6]

Over the past decade, the trend across regions and national income levels has tended to be toward heavier populations. India's level of 19 percent adults overweight is up from 14 percent in 2002 and 16 percent in 2005.[7] In Mexico

the figure has risen by 8 percentage points since 2002, while Brazil's is up by 7 points and the rate in the United Kingdom is up by 5 points.[8] East Asia has seen a 4 point increase over the period.[9]

Because wealthier people often have access to more and possibly less-healthy food and have less physically demanding work and more leisure time, it is often assumed that they tend to be heavier. And to some extent this relationship holds true: some 59 percent of adults in the 34 countries that belong to the Organisation for Economic Co-operation and Development (OECD) are overweight.[10] Worldwide, some 75 percent of the adults in the 10 richest countries are overweight, while in the 10 poorest, only 18 percent are.[11] (See Table 2.)

Indeed, looking at just world regions, being overweight appears to correlate reasonably well with income.[12] On average, people in low-income, tropical Africa have low BMIs, as do people in impoverished southern Asia. Middle-income Latin America and the Caribbean have middle-range BMIs, especially when the wealthier Southern Cone nations, whose incomes and BMIs are higher than in the rest of the region, are separated out. And at the high end of incomes, Europe generally has elevated levels, with more than half of the adult population overweight, while the United States posts the highest average BMI in the industrial world.[13]

At the national level, however, the situation is more complex. A comparison of percentages of people overweight in all countries and their GDPs reveals a positive but weak correlation.[14] Indeed, having more money for food and leisure does not necessarily mean people will become fat. Among the OECD nations, Japan has the lowest proportion of overweight adults (23 percent); the next is France (42 percent).[15] This is

Table 1. Overweight Adults Worldwide, 2002–10

Year	Overweight Adults
	(billion)
2002	1.454
2005	1.602
2010	1.934

Source: U.N. Population Division and WHO, Global Infobase.

Table 2. Overweight Adults in the 10 Richest and Poorest Countries

	Share of Adults Who Are Overweight
	(percent)
Richest Countries	
Qatar	62.7
Luxembourg	56.6
Singapore	25.4
Norway	51.5
Brunei	61.7
United States	78.6
Switzerland	57.7
Netherlands	48.2
Australia	71.1
Austria	59.1
Poorest Countries	
Malawi	14.0
Sierra Leone	37.7
Togo	28.2
Central African Rep.	4.9
Niger	21.1
Eritrea	18.1
Liberia	15.1
Zimbabwe	39.1
Burundi	33.7
Congo	10.8

Source: U.N. Population Division and WHO, Global Infobase.

far below the 70-plus percentages found in Australia and the United States.[16]

The 20 countries with the lowest BMI values are all poor nations in Asia and Africa. In fact, the first industrial nation with a low BMI value is Japan, ranked twenty-ninth.[17] The 20 nations with the highest rates of overweight people are literally all over the map, featuring countries as diverse as the United States, Mongolia, Bolivia, and Egypt.[18] Perhaps the greatest surprise is the diverse pattern of overweight found in Oceania. Melanesia, the bulk of whose people live in Papua New Guinea, exhibits a Third World pattern of relatively low BMI. At the other extreme, highly urbanized and developed Australia and New Zealand have rates close to North America. But by far the highest values are found in

Micronesia and Polynesia (two areas combined because data were available for so few countries).[19] There, nearly 88 percent of the population older than 15 is overweight.[20]

Clearly, then, standard of living and being overweight often show a relationship, but it is neither linear nor devoid of other influences, particularly urbanization and greater access to "junk" food, a trend that has been studied for some time.[21] Many studies have linked the U.S. rate with such factors as lack of exercise, overeating, poor dietary habits, and genetics, for example.[22] Genetic factors may play a part in Micronesia and Polynesia too, although urbanization, changing work patterns, and shifts in diet also seem to be involved.[23]

To further examine regional differences, the data can be broken down by gender. At the global level, women tend to be heavier than men. About 75 percent of overweight people are women.[24] But the percentages vary among nations. Table 3 lists the top 10 countries in which the percentage of adult men who are overweight exceeds that of adult overweight women.[25] Nearly all are industrial countries. In fact, with the exceptions of Cambodia and China, the top 30 countries in this category are industrial, with 23 of them in Europe.

Table 4 lists the opposite situation: the top 10 nations in which the percentage of adult women who are overweight exceeds that of adult overweight men.[26] Here the bias is strongly toward developing states. In the top 30, only Macedonia may not fit in that category.

In the 39 countries in which the numbers of overweight women and men are roughly equal (within ±5 percent), there is no obvious common social or economic thread; the United States, Ethiopia, Chile, the United Kingdom, and Bangladesh all fall into this category. Each case may depend on a unique set of local factors.

Local or cultural conditions may explain many national and local differences in whether higher percentages of men or women are overweight. Differing levels of physical work for different cohorts or the disproportionate deaths of some age-sex groups (from AIDS, for example)

Table 3. Top 10 Countries Where Share of Men Who Are Overweight Exceeds That of Women

Country	Share of Men Who Are Overweight	Share of Women Who Are Overweight	Difference
	(percent)		
Lithuania	62	44	18
Bulgaria	63	46	17
Estonia	51	34	17
Croatia	64	48	16
Italy	55	40	15
Greece	78	63	15
Denmark	55	41	14
Japan	30	16	14
China	45	32	13
Serbia	61	48	13

Source: U.N. Population Division and WHO, Global Infobase.

Table 4. Top 10 Countries Where Share of Women Who Are Overweight Exceeds That of Men

Country	Share of Men Who Are Overweight	Share of Women Who Are Overweight	Difference
	(percent)		
Lesotho	30	71	41
Haiti	19	58	39
Zimbabwe	17	51	34
Jamaica	47	79	32
South Africa	41	68	27
Morocco	31	58	27
Côte d'Ivoire	13	36	23
Mauritania	35	58	23
Sierra Leone	26	49	23
Gabon	30	52	22

Source: U.N. Population Division and WHO, Global Infobase.

are factors in some cases.[27] Also, since women in developing countries tend to be pregnant more often than they are in industrial nations, this may be contributing to that group's heaviness.[28]

Many of these factors could be occurring in southern Africa, where the percentage of overweight individuals is particularly high for a developing region, and the population of South Africa itself is becoming heavier.[29] A detailed study of one South African township found that the only strong correlation for overweight women was hunger as children, followed by adequate access to food as adults.[30] These conditions did not lead to men becoming heavier, and they may suggest why, worldwide, more women than men tend to be overweight.

What can be said is that a population with a higher income has more opportunity to become overweight. Whether this weight gain occurs depends on cultural and social norms, the habits of individuals and groups, and perhaps genetics as well. And when people become heavier, there is an increased risk of several medical problems. These include coronary heart disease, high blood pressure, stroke, diabetes, and gallstones.[31] But such problems do not always occur, and it is possible that with proper education and encouragement the worldwide trend toward being overweight can be reversed.

Richard H. Weil teaches general education at Brown College in Mendota Heights, Minnesota.

Notes

Oil Market Resumes Growth after Stumble in 2009 (pages 16–18)

1. BP, *Statistical Review of World Energy 2011: Historical Data*, available at www.bp.com, viewed 11 July 2011.
2. Ibid.
3. Ibid.
4. Ibid.
5. Ibid.
6. Ibid.
7. Ibid.
8. "OPEC Approves Biggest Cut Ever," *MSNBC.com*, 17 December 2008.
9. BP, op. cit. note 1.
10. Ibid.
11. Ibid.
12. Campbell Robertson, "Search Continues After Oil Rig Blast," *New York Times*, 21 April 2010.
13. Mark Clayton, "Offshore Drilling Ban: Will Revised Moratorium Appease Courts," *Christian Science Monitor*, 12 July 2010; James Kanter, "Europe Considers a Curb on Deepwater Drilling," *New York Times*, 12 October 2010.
14. National Commission on the BP Deepwater Horizon Oil Spill and Offshore Drilling, *Deep Water: The Gulf Oil Disaster and the Future of Offshore Drilling* (Washington, DC: January 2011), p. 216.
15. Ibid., p. 217; International Energy Agency (IEA), *Medium-Term Oil and Gas Markets 2010* (Paris: 2010), p. 15.
16. BP, op. cit. note 1.
17. Ibid.
18. Ibid.
19. Ibid.; IEA, "IEA Makes 60 Million Barrels of Oil Available to Market to Offset Libyan Disruption," press release (Paris: 23 June 2011); IEA, "IEA Collective Action—June 23, 2011: Frequently Asked Questions," at www.iea.org/files/faq.asp, viewed 7 July 2011.
20. IEA, "IEA Makes 60 Million Barrels of Oil Available," op. cit. note 19.
21. Ibid.
22. Wael Mahdi, "Saudis May Not Pump Intended 10 Million Barrels after IEA Move," *Bloomberg*, 28 June 2011; Rowena Mason, "OPEC Split Threatens Increase in Saudi Oil Production," (London) *Telegraph*, 29 June 2011; Benoit Faucon, "Iran Says Oil Output to Spark Price War If Not Matched by Demand," *Wall Street Journal*, 7 July 2011.
23. U.S. Department of Energy (DOE), Energy Information Administration (EIA), "Cushing, OK WTI Spot Price FOB (Dollars per Barrel)" and "Annual Cushing, OK WTI Spot Price FOB (Dollars per Barrel)," at www.eia.gov, both viewed 18 July 2011.
24. Ibid.
25. Worldwatch calculations based on BP, op. cit. note 1.
26. BP, op. cit. note 1.
27. Ibid.
28. Ibid.
29. Ibid.
30. Ibid.
31. Ibid.
32. IEA, *World Energy Outlook 2010* (Paris: 2010), p. 143.
33. BP, op. cit. note 1.
34. DOE, EIA, "Canada," at www.eia.gov/countries, viewed 13 July 2011.

Global Natural Gas Consumption Regains Momentum (pages 19–22)

1. BP, *Statistical Review of World Energy 2011*, available at www.bp.com, viewed 1 November 2011.
2. BP, op. cit. note 1. This figure reflects a share of primary energy use, defined to include oil, natural gas, coal, nuclear energy, hydroelectricity, and non-hydro renewables. In previous editions of the *Statistical Review of World Energy*, BP did not calculate primary energy use for non-hydro renewables. Consequently, the share of primary energy cited in the natural gas trend in *Vital Signs 2011* for 2009 was 23.8 percent. That figure would be 23.4 percent if non-hydro renewables were included.
3. BP, op. cit. note 1.
4. Ibid.
5. Ibid.
6. See, for example, International Energy Agency (IEA), *World Energy Outlook 2010* (Paris: 2010), p. 187. The IEA estimates remaining recoverable resources of conventional natural gas at the end of 2009 to be 404 trillion cubic meters and unconventional natural gas to be 380 trillion cubic meters.
7. U.S. Department of Energy (DOE), Energy Information Administration (EIA), and Advanced Resources International, *World Shale Gas Resource: An Initial Assessment of 14 Regions Outside the United States* (Washington, DC: 2011).
8. Roseanne Barrett, "Analysts' Money on Shale Gas," *The Australian*, 8 November 2011; "China Plans Subsidies to Tap Shale Gas Reserves Larger than U.S.," *Bloomberg Businessweek*, 20 October 2011; "Exxon to Start Fracking 2nd Polish Shale Gas Well," *Reuters*, 28 September 2011; "Exxon Mobil Looks for Shale Gas in Germany," *UPI.com*, 25 January 2011; Guy Chazan, "U.K. Gets Big Shale Find," *Wall Street Journal*, 22 September 2011.
9. BP, op. cit. note 1.
10. Ibid.
11. Ibid.
12. Ibid.
13. Ibid.
14. Ibid.
15. Ibid.; "Turkmenistan, Russia Agree to Resume Gas Supplies in 2010, Ending Impasse," *IHS Global Insight*, 23 December 2009.
16. BP, op. cit. note 1.
17. Ibid.
18. Ibid.
19. Ibid.
20. Ibid.
21. Ibid.
22. IEA, *Are We Entering a Golden Age of Gas? World Energy Outlook 2011 Special Report* (Paris: 2011), p. 15.
23. BP, op. cit. note 1.
24. Ibid.
25. IEA, op. cit. note 6, "Table 5.2: Primary Natural Gas Demand by Region in the New Policies Scenario," p. 182.
26. BP, op. cit. note 1.
27. Ibid.; Friends of the Earth, *Gas Flaring in Nigeria: A Human Rights, Environmental and Economic Monstrosity* (Amsterdam: Climate Justice Programme and Environmental Rights Action/Friends of the Earth Nigeria, 2005); World Bank, "Gas Flaring Reductions Avoid 30 Million Tons of Carbon Dioxide Emissions in 2010," press release (Washington, DC: 27 June 2011).
28. Michael F. Farina, *Flare Gas Reduction: Recent Global Trends and Policy Considerations* (General Electric, 2010).
29. World Bank, "Satellite Observations Show Declining Levels of Gas Flaring, Greenhouse Emissions," press release (Washington, DC: 17 November 2009); Farina, op. cit. note 28.
30. World Bank, "Russia, Kazakhstan Lead Way to Reduce Gas Flaring and Lower Emissions," press release (Washington, DC: 27 June 2011).
31. Farina, op. cit. note 28.
32. World Bank, op. cit. note 30.
33. BP, op. cit. note 1.
34. Ibid.
35. Ibid.
36. Ibid.
37. Anthony Melling, *Natural Gas Pricing and Its Future: Europe as the Battleground* (Washington, DC: Carnegie Endowment for International Peace, October 2010).
38. BP, op. cit. note 1.
39. Ibid.
40. Ibid.
41. Ibid.
42. Ibid.
43. International Gas Union (IGU), *World LNG Report 2010* (Oslo: 2011).
44. Ibid.
45. Ibid.
46. BP, op. cit. note 1.

47. Ibid.
48. Ibid.
49. Ibid.
50. Ibid.
51. DOE, EIA, "U.S. Natural Gas Imports Fall for Third Year in a Row," *Today in Energy*, 1 April 2011.
52. IGU, op. cit. note 43.
53. Ibid.
54. Ibid.
55. BP, op. cit. note 1.
56. Ibid.
57. Ibid.
58. Robert Cutler, "Turkmenistan Diversifies Gas Export Routes," *Central Asia-Caucasus Institute Analyst*, 8 July 2010.
59. David Blair, "BP Plans Gas Pipeline to Europe from Azerbaijan," *Financial Times*, 26 September 2011; "EU: Trans-Adriatic Pipeline 'Promising,'" *UPI.com*, 31 October 2011.
60. Vladimir Socor, "European Union Officially Endorses Trans-Caspian Pipeline to Link Up With Nabucco," *Eurasia Daily Monitor*, 20 September 2011; Peter Leonard, "Russia Slams EU Plans to Support Caspian Pipeline," *SFGate.com*, 13 September 2011.
61. DOE, EIA, "Libya Resumes Natural Gas Exports to Italy," *Today in Energy*, 20 October 2011.
62. John Daly, "Jordan Scrambling to Replace Lost Egyptian Natural Gas Imports," *OilPrice.com*, 27 October 2011.
63. "France Expands Nuclear Power Plans despite Fukushima," *BBC News*, 30 May 2011; Hiroko Tabuchi, "Cooling Problem Shuts Nuclear Reactor in Japan," *New York Times*, 4 October 2011.
64. Takeo Kumagei, "Japanese Utilities April-September LNG Purchases Reach Record High," *Platts*, 18 October 2011.

Nuclear Generation Capacity Falls (pages 23–25)

1. International Atomic Energy Agency (IAEA), "International Atomic Energy Agency Power Reactor Information System (IAEA-PRIS)," online database, at www.iaea.org/programmes/a2.
2. Ibid.
3. Ibid.
4. Ibid.
5. Ibid.
6. Ibid.
7. Ibid.
8. Ibid.
9. Ibid.
10. Mycle Schneider, Anthony Froggatt, and Steve Thomas, *World Nuclear Industry Status Report 2010–2011: Nuclear Power in a Post-Fukushima World* (Paris, Berlin, and Washington, DC: Worldwatch Institute, April 2011).
11. Électricité de France, "EDF Will Start Selling the First KWh Produced by the EPR at Flamanville in 2016," press release (Paris: 20 July 2011).
12. "'New Approach' Puts Back Flamanville 3," *World Nuclear News*, 21 July 2011.
13. IAEA, op. cit. note 1.
14. Ibid.
15. Ibid.
16. Ibid.
17. Ibid.
18. Ibid.
19. Ibid.
20. Schneider, Froggatt, and Thomas, op. cit. note 10. The average age of all decommissioned plants up until 1 April 2011 was taken from the *World Nuclear Industry Status Report*. The ages of the eight recently decommissioned plants in Germany are from www.world-nuclear.org/info/inf43.html and are incorporated into the average.
21. James Sullivan, "Will the Problems in Japan Lead to a Meltdown in the Nuclear Construction Industry?" *Value Line*, 20 July 2011.
22. Ibid.
23. Sebastian Ehreiser, "Country Perspective: United States," in Nina Netzer and Jochen Steinhilber, eds., *The End of Nuclear Energy? International Perspectives after Fukushima* (Berlin: Friedrich Ebert Stiftung, July 2011).
24. Ibid.
25. Ibid.
26. Ibid.
27. "China to Begin Approval Process for New Nuclear Power Plants," *Jagran Post*, 25 June 2011; Daniel Krahl and Su Junxia, "Country Perspective: China," in Netzer and Steinhilber, op. cit. note 23.
28. Krahl and Su, op. cit. note 27.
29. "China to Begin Approval Process," op. cit. note 27.
30. Krahl and Su, op. cit. note 27.
31. Hiroko Tabuchi, "Cooling Problem Shuts Nuclear Reactor in Japan," *New York Times*, 5 October 2011.
32. Ibid.
33. "Germany to Give Up Nuclear Power by 2022: Angela Merkel," *Jagran Post*, 10 June 2011.

34. Regine Günther, "Country Perspective: Germany," in Netzer and Steinhilber, op. cit. note 23; "Germany to Give Up Nuclear Power by 2022," op. cit. note 33.

35. "Swiss Cabinet Agrees to Phase Out Nuclear Power," *Reuters*, 25 May 2011.

36. Ibid.

37. Sullivan, op. cit. note 21.

38. Judy Dempsey and Sharon LaFraniere, "In Europe and China, Japan's Crisis Renews Fears About Nuclear Energy," *New York Times*, 17 March 2011.

39. Ibid.

40. "Contract Signed for Belarusian Reactors," *World Nuclear News*, 11 October 2011.

41. BP, *Statistical Review of World Energy June 2011*, online database, at www.bp.com.

42. Schneider, Froggatt, and Thomas, op. cit. note 10, p. 5; "World Net Nuclear Electric Power Generation (Billion Kilowatthours), 1980–2006" and "World Total Net Electricity Generation (Billion Kilowatt hours), 1980–2006," in U.S. Department of Energy, Energy Information Administration, *International Energy Annual 2006* (Washington, DC: 2006).

43. Schneider, Froggatt, and Thomas, op. cit. note 10.

44. Ibid.

45. Ibid.

Global Wind Power Growth Takes a Breather in 2010 (pages 26–28)

1. Global Wind Energy Council (GWEC), *Global Wind Report: Annual Market Update 2010* (Brussels: 2011).

2. Calculated from ibid.

3. Ibid.

4. Ibid.

5. Ibid.

6. Wu Oj, "China Takes Grid Connected Capacity to 31GW," *Windpower Monthly*, 17 January 2011.

7. Ibid.

8. Haibing Ma, "Beyond the Numbers: A Closer Look at China's Wind Power Success," *ReVolt*, 28 February 2011.

9. "Wind Power: China Finds Itself Awash in Wind Turbine Factories," *ClimateWire*, 12 March 2010.

10. "China Grids to Connect 90 m kW of Wind Power by 2015," *China Daily*, 16 April 2011.

11. Ibid.

12. Government of China, at www.gov.cn/2011lh/content_1825838_4.htm (in Chinese).

13. GWEC, op. cit. note 1, p. 12.

14. GWEC, "Latest News: Global Wind Capacity Increases by 22% in 2010," 2 February 2011.

15. U.S. Department of Energy, Energy Information Agency, "Table 1.17.B, Net Generation from Wind by State by Sector," at www.eia.gov, viewed 19 March 2011.

16. Ibid.

17. GWEC, "United States," at www.gwec.net/index.php?id=121, viewed 22 April 2011.

18. European Wind Energy Association (EWEA), *Wind in Power: 2010 European Statistics* (Brussels: February 2011).

19. Ibid., p. 9.

20. Ibid., p. 5.

21. Ibid.

22. Ibid.

23. GWEC, op. cit. note 1.

24. Indian Wind Energy Association, "Installed Capacity Per State (MW)," at www.inwea.org, viewed 2 April 2011.

25. GWEC, "India," at www.gwec.net, viewed 27 March 2011.

26. GWEC, op. cit. note 1, p. 11.

27. Ibid., p. 16.

28. Ibid.

29. EWEA, *The European Offshore Wind Industry: Key Trends and Statistics 2010* (Brussels: January 2011).

30. Ibid.

31. Business Insights Ltd., *The Onshore and Offshore Wind Power Generation Outlook* (London: November 2010), p. 142.

32. Ibid.

33. "New Douglas-Westwood Research Forecasts Total Offshore Wind Spend of €38 Billion," *Windtech International*, 22 November 2010.

34. Business Insights Ltd., op. cit. note 31.

35. "Gamesa, Northrop Grumman Shipbuilding Launch Offshore Wind Technology Center," *OffshoreWIND.biz*, 11 February 2011.

36. "Wind Turbine Prices Fall to Their Lowest in Recent Years," *Bloomberg New Energy Finance*, 7 February 2011.

37. Ibid.

38. "Clean Energy Investment Storms to New Record in 2010," *Bloomberg New Energy Finance*, 11 January 2011.

39. Ibid.

40. "Vestas Retains Pole Position as Top Wind Turbine Manufacturer," *NewNet.com*, 16 March 2011.

Another Record Year for Solar Power, But Clouds on the Horizon (pages 29–31)

1. Worldwatch estimate based on IMS Research, "PV Installations Reached 17.5 GW in 2010; New Report Forecasts 20.5 GW for 2011," press release (Wellingborough, U.K.: 17 January 2011).

2. Yingling Liu,"Solar Power Experiences Strongest Year of Growth Yet," in Worldwatch Institute, *Vital Signs 2010* (Washington, DC: 2010).

3. European Photovoltaic Industry Association (EPIA), *Solar Generation 6* (Brussels: February 2011); Worldwatch calculation based on European household electricity consumption of 3.94 megawatt-hours per year, derived from European Commission, *Eurostat*, at epp.eurostat.ec.europa.eu, and an average solar capacity factor of 15 percent.

4. REN21, *Renewables: 2011 Global Status Report* (Paris: 2011), p. 25; Worldwatch calculation based on concentrated solar power project database of the National Renewable Energy Laboratory (NREL), at www.nrel.gov/csp/solarpaces.

5. Germany from Federal Ministry for the Environment, Nature Conservation and Nuclear Safety, "Renewables' Contribution to Energy Supply in Germany Continued to Rise in 2010," at www.erneuerbare-energien.de, 16 March 2011; Spain is Worldwatch calculation based on Red Eléctrica de España, *El Sistema Electric Español. Avance del Informe 2010* (Madrid: December 2010), p. 7.

6. EPIA, "Solar Photovoltaics: 2010 a Record Year in All Respects," press release (Brussels: 22 February 2011).

7. European Union wind installations totaled 16.3 GW in 2010: Mark Konold, "Global Wind Power Takes a Breather in 2010," *Vital Signs Online*, Worldwatch Institute, 30 June 2011.

8. Data from German Federal Network Agency (Bundesnetzagentur), "Solar Expansion in 2010 Totalling Capacity of 7,400 MW is Almost Twice as High as 2009," press release (Bonn: 21 March 2011).

9. IMS Research op. cit. note 1; James Russell, "Record Growth in Photovoltaic Capacity and Momentum Builds for Concentrating Solar Power," in Worldwatch Institute, *Vital Signs 2011* (Washington, DC: 2011).

10. German Federal Network Agency, op. cit. note 8.

11. REN21 op. cit. note 4, p. 23.

12. Ibid.

13. iSuppli, *PV Installations, Systems, and Inverters: What is Ahead for 2011?* (El Segundo, CA: February 2011).

14. Data from Commissariat Général au Développement Durable, *Tableau de Bord Éolien-photovoltaïque: Quatrième Trimestre 2010* (Paris: February 2011).

15. "La Energíafotovoltaicaprodujo un 5,7% Más en 2010, Siendo C-LM la de Mayor Potencia, con 853 MW," *Eleconomista.es*, 7 April 2011.

16. iSuppli, op. cit. note 13.

17. Ibid.; only 878 MW is grid-connected, according to Solar Energy Industries Association (SEIA) and Greentech Media (GTM), *U.S. Solar Market Insight: 2010 Year-In-Review* (Washington, DC, and Boston: March 2011).

18. Russell, op. cit. note 9.

19. NREL, op. cit. note 4.

20. Ibid.

21. SEIA and GTM, op. cit. note 17.

22. NREL, op. cit. note 4.

23. "26th Annual Data Collection Results: Another Bumper Year for Manufacturing Masks Turmoil," *PV News*, May 2010; "27th Annual Data Collection Results," *PV News*, May 2011.

24. Solarbuzz, "Solarbuzz Reports World Solar Photovoltaic Market Grew to 18.2 Gigawatts in 2010, Up 139% Y/Y," press release (New York: 15 March 2011).

25. "27th Annual Data Collection Results," op. cit. note 23.

26. Ibid.

27. Ibid.; IMS Research, "Asian Electronics Giants Aim for Big Slice of PV Pie," press release (Wellingborough, U.K.: 17 February 2011).

28. SunTech, "Suntech Reports Fourth Quarter and Full Year 2010 Financial Results," press release (Wuxi, China: 8 March 2011); JA Solar, Corporate Overview, at www.jasolar.com; "PHOTON: Global PV Cell Production Expanded 118% to 27.2 GW in 2010," Solar Server, 25 March 2011, at www.solarserver.com.

29. Mario Ragwitz et al., *Recent Experiences with Feed-in Tariff Systems in the EU—A Research Paper for the International Feed-In Cooperation* (Karlsruhe, Germany: November 2010).

30. "Spain Slashes Prices 45% for New Ground-Based Solar Plants, 5% for Homes," *Bloomberg*, 19 November 2010.

31. "Czech President Approves Controversial Solar Tax," *PV Magazine*, 16 December 2010.

32. "EEG Amendment Agreed: German Bundestag Concludes on the Phasing Out of Nuclear Energy by 2022," *Solarplaza*, 5 July 2011.

33. "27th Annual Data Collection Results," op. cit. note 23.

34. Ibid.

35. Ibid.; Tier 1 producers are firms deemed by lenders to be the most bankable or to have the lowest risk profile in regards to module performance and ability for the producer to honor their warranties (GTM Research Note, "Module Producer Bankability Assessment").

36. Syanne Olson, "Lux Research Report Examines Module Cost Structure Breakdown, Weighs c-Si Against Other Technology," *PV Tech*, 17 November 2010; "27th Annual Data Collection Results," op. cit. note 23.

37. "27th Annual Data Collection Results," op. cit. note 23.

38. Ibid.

39. Ibid.

40. G. Barbose, N. Darghouth, and R. Wiser, *Tracking the Sun III: The Installed Costs of Photovoltaics from 1998-2009* (Berkeley, CA; Lawrence Berkeley National Laboratory, December 2010).

41. Ibid.

42. "China's Self-Sustaining Solar Economy," *PV News*, February 2011.

43. UBS Investment Research, *Global Solar Industry— Outlook 2011* (30 November 2010); Deutsche Bank, Global Markets Research, *Solar Photovoltaic Industry 2011 Outlook: FIT Cuts in Key Markets Point to Oversupply* (5 January 2011).

44. UBS Investment Research, op. cit. note 43; Deutsche Bank, op. cit. note 43.

Biofuels Regain Momentum (pages 32–34)

1. REN21, *Renewables 2011 Global Status Report* (Paris: 2011), p. 32.

2. Ibid.

3. BP, *BP Statistical Review of World Energy* (London: June 2011), p. 3; REN21, op. cit. note 1, p. 31.

4. REN21, op. cit. note 1.

5. Ibid.

6. Ibid., p. 31.

7. Ibid., p. 32.

8. Ibid., p. 75.

9. Ibid.

10. Ibid., p. 31.

11. U.S. Environmental Protection Agency, "E15 (A Blend of Gasoline and up to 15% Alcohol)," at www.epa.gov/otaq.

12. REN21, op. cit. note 1, p. 31; Renewable Fuels Association, *2011 Ethanol Industry Outlook* (Washington, DC: February 2011), p. 27.

13. REN21, op. cit. note 1, p. 31.

14. Ibid., p. 32.

15. Ibid.

16. Ibid., p. 31.

17. Ibid., p. 32.

18. Ibid.

19. Ibid., p. 75; REN21, *Renewables 2010 Global Status Report* (Paris: 2010), p. 56.

20. REN21, op. cit. note 1.

21. Ibid., p. 32.

22. Rohan Boyle, "Latest Developments in Solar, Wind, Biofuels and Marine," *Bloomberg New Energy Finance*, Monthly Briefing, March 2010, p. 6.

23. REN21, op. cit. note 1, p. 46.

24. Michael McConnell, Erik Dohlman, and Stephen Haley, "World Sugar Price Volatility Intensified by Market and Policy Factors," *Amber Waves* (U.S. Department of Agriculture (USDA)), September 2010.

25. European Biodiesel Board, "2009–2010: EU Biodiesel Industry Restrained Growth in Challenging Times," press release (Brussels: 22 July 2010).

26. Charlie Dunmore, "Climate Impact Threatens Biodiesel Future in EU," *Reuters*, 8 July 2011.

27. Ibid.

28. Jim Lane, "Biofuels Mandates Around the World," *Biofuels Digest*, 21 July 2011.

29. Shane Romig, "Despite Global Hiccups, Argentine Biodiesel Booms," *MarketWatch*, 18 April 2011.

30. Ibid.; Bryan Sims, "Argentina Increases Biodiesel Mandate," *Biodiesel Magazine*, 13 July 2010 (converted from tons to gallons).

31. Sims, op. cit. note 30; Lane, op. cit. note 28.

32. Romig, op. cit. note 29.

33. Lane, op. cit. note 28.

34. Ibid.

35. Ibid.; REN21, op. cit. note 1, p. 86.

36. Nuel Navarrete, "E.P.A. Lowers Cellulosic Ethanol Requirement for 2011," *EcoSeed*, 1 December 2010.

37. Ibid.

38. International Energy Agency, *Sustainable Production of Second-Generation Biofuels* (Paris: 2010), pp. 22–23; Paul Adler et al., "Life Cycle Assessment of Net Greenhouse-Gas Flux for Bioenergy Cropping Systems," *Ecological Applications*, vol. 17 (2007), pp. 675–91; Jorn Scharlemann and William Laurance, "How Green Are Biofuels?" *Science*, 4 January 2008, pp. 43–44; David Tilman et al., "Carbon-Negative Biofuels from Low-Input High-Diversity Grassland Biomass," *Science*, 8 December 2006, pp. 1,598–1,600.

39. Navarrete, op. cit. note 36.

40. Gavin Maguire, "USDA Report May be Bad for Ethanol," *Reuters*, 12 July 2010; USDA, Economic Research Service, "Corn: Background," updated 18 February 2009, at www.ers.usda.gov/Briefing/Corn/background.htm.

41. Timothy Gardner, "US Senate Deal Would Axe $6 Bln Ethanol Tax Credit," *Reuters*, 7 July 2011.

42. Carey Gillam, "Ethanol Grown Up, Will Withstand U.S. Subsidy Loss," *Reuters*, 17 June 2011.

43. Renewable Fuels Association, "Statistics," at www.ethanolrfa.org/pages/statistics.

44. American Coalition for Ethanol, "Federal Legislation," at www.ethanol.org/index.php?id=78.

45. Peter Murphy, "Low Yields Put Dent in Brazil Sugar, Ethanol – Unica," *Reuters*, 7 July 2011.

46. Michael Grunwald, "The Clean Energy Scam," *Time*, 27 March 2008.

47. Katia Cortes, "Brazil to Need $550 Billion Energy Investment by 2019," *Bloomberg Businessweek*, 29 November 2010.

Global Hydropower Installed Capacity and Use Increase (pages 35–37)

1. BP, *Statistical Review of World Energy* (London: June 2011).

2. Ibid.

3. Ibid.

4. Ibid.; REN21, *Renewables 2011 Global Status Report* (Paris: 2011).

5. BP, op. cit. note 1.

6. Ibid.

7. REN21, op. cit. note 4.

8. Ibid.

9. Ibid.

10. BP, op. cit. note 1.

11. Ibid.

12. Ibid.

13. REN21, op. cit. note 4.

14. Ibid.

15. Ibid.

16. Ibid.

17. Ibid.

18. Ibid.

19. Ibid.

20. Ibid.

21. "List of World's Largest Hydroelectricity Plants and Countries—China Leading in Building Hydroelectricity Stations," *Green World Investor*, 29 March 2011.

22. Ibid.

23. Ibid.

24. BP, op. cit. note 1.

25. Ibid.

26. Ibid.

27. U.S. Department of Energy (DOE), Energy Information Administration (EIA), *International Energy Statistics* (Washington DC: 2011).

28. Ibid.

29. BP, op. cit. note 1.

30. REN21, op. cit. note 4.

31. REN21, *Renewables 2010 Global Status Report* (Paris: 2010).

32. Ibid.

33. REN21, op. cit. note 4.

34. Ibid.

35. Ibid.

36. Ibid.

37. Ibid.

38. DOE, EIA, "BPA Curtails Wind Power Generators during High Hydropower Conditions," *Today in Energy*, 15 June 2011.

39. Ibid.

40. Ibid.

41. Ibid.

42. REN21, op. cit. note 31.

43. Ibid.

44. Ibid.

Energy Poverty Remains a Global Challenge for the Future (pages 38–41)

1. United Nations, "International Year for Sustainable Energy for All," at www.un.org/en/events; General Assembly resolution at www.un.org/ga.

2. REN21, *Renewables 2011 Global Status Report 4* (Paris: 2011), p. 66.

3. U.N. Development Programme (UNDP), *Delivering Energy Services for Poverty Reduction: Success Stories from Asia and the Pacific* (Bangkok: 2008), p. 2.

4. Oxford Poverty & Human Development Initiative, *Measuring Energy Poverty: Focusing on What Matters* (Oxford: 2010), pp. 4–5.

5. International Energy Agency (IEA), *Energy and Development Methodology* (Paris: 2011), p. 3.

6. U.N. Industrial Development Organization, *Energy*

Access: Time for Action (Durban: 2011), p. 3.

7. IEA, op. cit. note 5, p. 1.

8. Figure of 1.3 billion from IEA, "Access to Electricity," World Energy Outlook Web site, at www.iea.org/weo.

9. Unreliable access figure from Advisory Group on Energy and Climate Change (AGECC), *Summary Report and Recommendations* (New York: United Nations, 2010), p. 7.

10. For lack of access to modern fuels, the lower bound is from IEA, op. cit. note 8; higher end is from UNDP and World Health Organization (WHO), *The Energy Access Situation in Developing Countries: A Review Focusing on the Least Developed Countries and Sub-Saharan Africa* (New York: 2009), p. 1. Figure 1 is based these three sources.

11. Table 1 is adapted from IEA, op. cit. note 8.

12. IEA, op. cit. note 8.

13. U.N. Environment Programme (UNEP), *Towards a Green Economy* (Paris: 2011), p. 208.

14. IEA, op. cit. note 8.

15. Figure 2 adapted from ibid.

16. Projection for 2015 from Lighting Africa, "Solar Lighting for the Base of the Pyramid—Overview of an Emerging Market," Executive Summary, at www.lightingafrica.org/component/docman/doc_download/153-executive-summary-final.html; 2030 from Lighting Africa, "Landmark Conference Aims to Transform Low Cost Off-grid Lighting Sector," press release (Nairobi: 18 May 2010).

17. IEA, op. cit. note 8.

18. Ibid.

19. Ibid.

20. Ibid.

21. Ibid.

22. Ibid.

23. Ibid.

24. Ibid.

25. U.N. Department of Economic and Social Affairs, *World Economic and Social Survey 2011: The Great Green Technological Transformation* (New York: 2011), p. 36.

26. AGECC, op. cit. note 9, pp. 14–16.

27. Ministry of Mines and Energy (Brazil), *Light for All. A Historic Landmark: 10 Million Brazilians out of the Darkness* (Brasilia: undated); Alexandra Niez, *Comparative Study on Rural Electrification Policies in Emerging Economies: Keys to Successful Policies* (Paris: IEA, 2010), pp. 23, 26.

28. UNEP, "Traditional Use of Biomass," at www.unep.org.

29. UNDP and WHO, op. cit. note 10, p.2.

30. Ibid.

31. IEA, *Energy for All. Financing Access for the Poor*, special early excerpt of the *World Energy Outlook 2011* (Paris: 2011), p. 11.

32. Ibid.

33. Ibid.

34. Ibid.

35. Table 2 from ibid.

36. Ibid., p. 7.

37. Ibid.

38. Ibid.

39. Ibid.

40. Ibid.

41. Ibid.

42. Ibid., pp. 21–24.

43. UNDP and WHO, op. cit. note 10, p. 101.

44. Ibid.

45. UNEP, op. cit. note 13, pp. 208–09.

46. REN 21, op. cit. note 2, p. 33.

47. AGECC, op. cit. note 9, p. 17.

48. Ibid.

49. UNDP and WHO, op. cit. note 10, pp.19–20.

50. Ibid.

Auto Industry Stages Comeback from Near-Death Experience (pages 44–47)

1. Colin Couchman, IHS Automotive, London, e-mail to author, 31 May 2011.

2. Ibid.

3. Ibid.

4. International Organization of Motor Vehicle Manufacturers, "World Motor Vehicle Production by Country and Type," at oica.net/wp-content/uploads/all-vehicles-2010-provisional.pdf, viewed 24 May 2011.

5. "2011: Recover, Rebalance, and Rebound," *PWC Autofacts Quarterly Forecast Update*, January 2011, p. 5.

6. Couchman, op. cit. note 1.

7. Ibid.

8. Ibid.

9. Ibid.

10. Carlos Gomes, *Global Auto Report* (Scotiabank Group), 29 March 2011, p. 1.

11. Ibid., p. 2.

12. Hiroko Tabuchi, "Toyota Says No Full Production Until Year's End," *New York Times*, 22 April 2011.

13. Couchman, op. cit. note 1.

14. Ibid.
15. "2011: Recover, Rebalance, and Rebound," op. cit. note 5, p. 5.
16. Ibid.
17. Ibid.
18. Ibid.
19. Couchman, op. cit. note 1.
20. Carlos Gomes, *Global Auto Report* (Scotiabank Group), 5 January 2011, p. 1.
21. Ibid., p. 2.
22. "2011: Recover, Rebalance, and Rebound," op. cit. note 5, p. 3.
23. Patti Waldmeir, "China Car Sales Stay in the Fast Lane," *Financial Times*, 4 January 2011; "License Quotas to Control Gridlock in Beijing," *China Daily*, 23 December 2010.
24. Michael Wines, "Multiplying Drivers Run Over Beijing Traffic Plan," *New York Times*, 22 December 2010.
25. "2011: Recover, Rebalance, and Rebound," op. cit. note 5, p. 3.
26. Ibid., p. 4.
27. Ibid.
28. Couchman, op. cit. note 1.
29. Ibid.
30. Brazil from "2011: Recover, Rebalance, and Rebound," op. cit. note 5, p. 4; China and India from Gomes, op. cit. note 20, p. 2.
31. IHS Automotive, PWC Autofacts, the Global Auto Report, and other sources like the International Organization of Motor Vehicle Manufacturers all draw somewhat different boundaries and thus offer production statistics that are not fully compatible with each other.
32. "2011: Recover, Rebalance, and Rebound," op. cit. note 5, pp. 6–7.
33. Ibid., p. 6.
34. Ibid.
35. Zhongxiu Zhao and Zhi Lv, "Global Supply Chain and the Chinese Auto Industry," *The Chinese Economy*, November-December 2009, p. 31.
36. Ibid.
37. "Automärkte: Chinas Autobauer auf mühsamer Aufholjagd," *Die Zeit* (Germany), 25 April 2011.
38. Timothy J. Sturgeon et al., "Globalisation of the Automotive Industry: Main Features and Trends," *International Journal of Technological Learning, Innovation and Development*, vol. 2, nos 1/2 (2009), pp. 13–14.
39. International Council on Clean Transportation (ICCT), "Global Passenger Vehicle Standards," at www.theicct.org, viewed 26 May 2011.
40. Ibid.
41. Ibid.
42. Ibid.
43. ICCT, "Datasheet of Global Passenger Vehicle FE/GHG Regulations," at www.theicct.org/info/data/Global_PV_Std_Jan2011Update_datasheet.xlsx.
44. Ibid.
45. Ibid.
46. Marc Cooper, *Gasoline Prices and Expenditures in 2011* (Washington, DC: Consumer Federation of America, March 2011).
47. Calculated from U.S. Environmental Protection Agency (EPA), *Light-Duty Automotive Technology, Carbon Dioxide Emissions, and Fuel Economy Trends: 1975 Through 2010* (Washington, DC: November 2010), Appendix C. The data are for "adjusted composite" fuel economy results—that is, combined city/highway laboratory drive cycles that are adjusted to approximate real-world conditions and vehicle usage.
48. Ibid., p. v.
49. Ibid., Table 3, p. 22.
50. Japan from Suzanne Ashe, "U.S. Car Market Lags behind Europe, Japan in 'Green' Fuel," *The Car Tech blog, CNET Reviews*, 17 June 2010; European data from European Federation for Transport and Environment (T&E), *How Clean Are Europe's Cars? An Analysis of Carmaker Progress towards EU CO2 Targets in 2009* (Brussels: November 2010), p. 22.
51. T&E, op. cit. note 50.
52. EPA, op. cit. note 47, Table 5, p. 28; T&E, op. cit. note 50, p. 22.
53. EPA, op. cit. note 47.
54. Ibid.; T&E, op. cit. note 50.
55. "Hybrid Car Statistics," *All-electric-vehicles.com*, undated, at www.all-electric-vehicles.com.
56. Toyota, "Sales in Japan of TMC Hybrids Top 1 Million Units," press release (Toyota City, Japan: 5 August 2010).
57. "Hybrid Car Statistics," op. cit. note 55.
58. Green Car Congress, "Cumulative Worldwide Sales of Toyota Hybrids Top 3M Units," 8 March 2011, at www.greencarcongress.com.
59. Sales in 2010 from Toyota, "Worldwide Sales of TMC Hybrids Top 3 Million Units," press release (Toyota City, Japan: 8 March 2011).
60. "Toyota to Double Hybrid Output in 2011: Report," *Reuters*, 18 January 2010.

61. Zhang Qi, "Energy Plan to Focus on Cars," *China Daily*, 30 December 2010.
62. Ibid.
63. "Charging Forward: Electric Vehicle Survey," *PWC Autofacts*, April 2011, at www.pwc.com/en_GX/gx/automotive/pdf/charging-forward-electric-vehicle-survey.pdf.
64. Ibid.
65. Ghosn cited in Alan Ohnsman, "Hybrid, Battery Car Demand Overhyped, J.D. Power Says," *Bloomberg*, 27 October 2010.
66. HSBC Global Research, *Sizing the Climate Economy* (London: September 2010), p. 20.
67. Ohnsman, op. cit. note 65.
68. J.D. Power and Associates, "Despite Rising Fuel Prices, the Outlook for 'Green' Vehicles Remains Limited for the Foreseeable Future," press release (Westlake Village, CA: 27 April 2011).
69. T&E, *How to Avoid an Electric Shock. Electric Cars: From Hype to Reality* (Brussels: November 2009).
70. Ibid.

High-Speed Rail Networks Expand (pages 48–51)

1. International Union of Railways (UIC), *Km of High Speed Lines in the World*, at www.uic.org/IMG/pdf/20110701_b1_resume_km_of_hs_lines_in_the_world.pdf.
2. UIC, "General Definitions of Highspeed," at www.uic.org, last updated 23 July 2010.
3. "China Poised to Become World's High-Speed Rail Leader," *China Daily*, 20 September 2009.
4. UIC, op. cit. note 1.
5. Ibid.
6. Calculated from total track length data in UIC, *Railway Statistics 2010 Synopsis* (Paris: 2011).
7. UIC, op. cit. note 1.
8. Ibid.
9. Ibid.
10. Ibid.
11. Ibid.
12. American Public Transportation Association (APTA), *High-Speed Rail Investment. Background Data* (Washington, DC: February 2011), p. 26.
13. UIC, "World High Speed Rolling Stock," 20 January 2011, at www.uic.org.
14. Ibid.
15. Ibid.
16. SCI Verkehr, *The Worldwide Market for Railway Technology 2009–2013* (Cologne: September 2008), p. 8;

"Schnelle Züge: Das Rennen zwischen ICE, TGV, und Co.," *Manager Magazin*, 28 July 2010.
17. UIC, *Rail and Sustainable Development* (Paris: April 2011), p. 7.
18. UIC, op. cit. note 6.
19. Calculated from ibid.
20. Klaus Ebeling, "40 Years of High-speed Railways. High-speed Railways in Germany," *Japan Railway & Transport Review*, March 2005, p. 36.
21. Christopher P. Hood, *Shinkansen: From Bullet Train to Symbol of Modern Japan* (New York: Routledge, 2006), p. 214.
22. Central Japan Railway Company, *The Number of Train Departures and Ridership*, at english.jr-central.co.jp (figure is for 2008); comparison from Yonah Freemark, "Meet The Train Makers, Part 4: The Japanese," *The Infrastructurist*, 10 November 2009.
23. Katsuhiro Yamaguchi and Kiyoshi Yamasaki, *High-Speed Inter-City Transport System in Japan: Past, Present, and Future*, Discussion Paper 2009-17 (Paris: Organisation for Economic Co-operation and Development and International Transport Forum, Joint Transport Research Centre, December 2009), p. 16.
24. "TGV High-Speed Rail Network, France," *Railway-Technology.com*, at www.railway-technology.com/projects/frenchtgv.
25. European Commission, Directorate-General for Energy and Transport, *EU Energy and Transport in Figures. Statistical Pocketbook 2010* (Luxembourg: Office for Official Publications of the European Communities, 2010), pp. 123–24.
26. UIC, "High Speed Rail Traffic in Europe 2000–2009," at www.uic.org, last updated 20 July 2010.
27. European Commission, op. cit. note 25; UIC, op. cit. note 26.
28. European Commission, op. cit. note 25, p. 124.
29. Yonah Freemark, "Getting the Price Right: How Much Should High-Speed Fares Cost?" *TheTransportPolitic.com*, 8 September 2009.
30. Ministère de l'Écologie, de l'Énergie, du Développement durable et de la Mer, Direction Générale des Infrastructures, des Transports et de la Mer, *Avant-Projet. Schéma national des infrastructures de transport soumis à concertation* (Paris: July 2010), p. 57.
31. Ibid.
32. "High-Speed Railways in Spain," New Technologies in Spain Series, *Technology Review* (Cambridge, MA: Massachusetts Institute of Technology, 2011), p. 2; W. E. Romp and J. Oosterhaven, *Indirect Economic Effects of a Rail Link Along the Afsluitdijk* (Amsterdam:

University of Amsterdam, Amsterdam School of Economics, January 2002).

33. Morgan Stanley Research, *China High-Speed Rail. On the Economic Fast Track*, Blue Paper (New York: 15 May 2011), p. 4.

34. UIC, op. cit. note 1.

35. Giles Tremlett, "Spain's High-speed Trains Win Over Fed-up Flyers," (London) *Guardian*, 13 January 2009.

36. "China to Have 13,000 KM High-speed Rail in Operation in 2011,"*Asia Today*, 16 March 2011.

37. Morgan Stanley Research, op. cit. note 33, p. 47.

38. APTA, op. cit. note 12, pp. 25–26; Jamil Anderlini, "Fresh Blow for China's High-speed Railway," *Financial Times*, 23 March 2011.

39. Michael Wines and Keith Bradsher, "China Rail Chief's Firing Hints at Trouble," *New York Times*, 17 February 2011.

40. Mure Dickie, "China Puts the Brakes on High-speed Rail," *Financial Times*, 28 June 2011.

41. Michael Wines and Sharon LaFraniere, "In Baring Facts of Train Crash, Blogs Erode China Censorship," *New York Times*, 28 July 2011.

42. Simon Rabinovitch, "China Suspends New High Speed Rail Plans," *Financial Times*, 11 August 2011.

43. "Das ICE-Unglück von Eschede," *Norddeutscher Rundfunk* (Germany), at www.ndr.de, viewed 30 August 2011.

44. "The Evolving Shinkansen. A Crystallization of Cutting-Edge Technology," *Trends in Japan*, at web -japan.org.

45. UIC, op. cit. note 13.

46. Central Japan Railway Company, *Data Book 2010* (Nagoya, Japan: 2010), p. 6.

47. Ibid., p. 14.

48. APTA, op. cit. note 12, p. 26.

49. Ibid.

50. Ibid., p. 27.

51. Ibid.

52. Morgan Stanley Research, op. cit. note 33, pp. 53–54.

53. Reinhard Clever and Mark M. Hansen, "Interaction of Air and High-Speed Rail in Japan," *Transportation Research Record*, No. 2043 (Washington, DC: Transportation Research Board of the National Academies, 2008), p. 2.

54. Ibid.

55. Ibid.

56. APTA, op. cit. note 12, p. 32. The source reports emissions in terms of pounds per passenger mile. Conversion by author.

57. Ibid.

58. International Council on Clean Transportation, "Datasheet of Global Passenger Vehicle FE/GHG Regulations," at www.theicct.org/info/data/Global _PV_Std_Jan2011Update_datasheet.xlsx.

59. Ibid.

60. Central Japan Railway Company, op. cit. note 46, p. 15.

61. APTA, op. cit. note 12, p. 35.

Carbon Markets Struggle to Maintain Momentum (pages 54–57)

1. World Bank, *State and Trends of the Carbon Market 2010* (Washington, DC: May 2010).

2. Ibid.

3. Ibid.

4. Christina Hood, *Reviewing Existing and Proposed Emissions Trading Systems* (Paris: Organisation for Economic Co-operation and Development/International Energy Agency, November 2010).

5. A. Denny Ellerman and Barbara K. Buchner, "Overallocation or Abatement? A Preliminary Analysis of the EU ETS Based on the 2005–06 Emissions Data," *Environmental and Resource Economics*, vol. 41, no. 2 (2008), pp. 267–87.

6. Hood, op. cit. note 4.

7. World Bank, op. cit. note 1.

8. Ibid.

9. Ibid.

10. Hood, op. cit. note 4.

11. Norwegian Mission to the European Union, "Norway to Join EU ETS," 30 October 2007.

12. Hood, op. cit. note 4.

13. Ibid.

14. Ibid.

15. Ibid.

16. World Bank, op. cit. note 1.

17. Ibid.

18. Ibid.

19. Ibid.

20. Hood, op. cit. note 4.

21. World Bank, op. cit. note 1.

22. State of California, *Statement of Vote: November 2, 2010, General Election*, revised 6 January 2011.

23. Zhang Xiang, "China Sets Emission Cuts as 'Binding Goals' in 2011–15 Period," *Xinhua News*, 27 October 2010.

24. David Stanway, "China Regions to Have Binding CO2 Targets," *Reuters*, 12 January 2011.

25. John M. Broder, "'Cap and Trade' Loses Its Standing

as Energy Policy of Choice," *New York Times*, 25 March 2010.

26. World Bank, op. cit. note 1.

27. "East Asian Cap and Trade Plans Hit the Wall," *carbonpositive.net*, 18 January 2011.

28. Ibid.

29. Audrey Young, "Govt May Ditch Emissions Trading Scheme," *New Zealand Herald*, 29 April 2010.

30. "East Asian Cap and Trade Plans Hit the Wall," op. cit. note 27.

31. "Italy Joins French Calls for EU Carbon Tariff," *euractive.com*, 19 April 2010.

32. United Nations Framework Convention on Climate Change, available at unfccc.int/kyoto_protocol.

33. World Bank, op. cit. note 1.

34. Ibid.

35. Worldwatch calculations based on ibid.

36. World Bank, op. cit. note 1.

37. Ibid.

38. Ibid.

39. Katherine Hamilton et al., *Building Bridges: State of the Voluntary Carbon*, 2010 ed. (Washington, DC: Ecosystem Marketplace, 2010).

40. Ibid.

41. Lee Barken, "New Life for REDD in California Compliance Market," *Ecosystem Marketplace*, 5 August 2010.

Carbon Capture and Storage Attracts Government Attention (pages 58–60)

1. Global CCS Institute, *The Global Status of CCS: 2010* (Canberra, Australia: 2011).

2. Ibid.

3. Global CCS Institute, "Status of CCS Project Database," 29 March 2011, at www.globalccsinstitute.com, viewed 5 April 2011. The Global CCS Institute defines "large" projects as those that sequester at least 1 Mtpa of CO_2, or at least 0.5 Mtpa of CO_2 in the case of natural gas processing projects. Its database of large integrated CCS projects includes plants in five stages of development: Identify, Evaluate, Define, Execute, and Operate. The eight plants referred to here are all in the operate stage.

4. Total storage from Global CCS Institute, op. cit. note 3; vehicle equivalence is a Worldwatch calculation based on U.S. Environmental Protection Agency, "Passenger Vehicle Conversion," at www.epa.gov/cleanenergy/energy-resources/refs.html#vehicles.

5. Global CCS Institute, op. cit. note 3.

6. Bert Metz et al., *Carbon Dioxide Capture and Storage* (Cambridge, U.K.: Cambridge University Press, for Intergovernmental Panel on Climate Change, 2005).

7. Global CCS Institute, op. cit. note 1, p. 23.

8. Data on large-scale projects from Global CCS Institute, op. cit. note 3; pilot project information from Massachusetts Institute of Technology, "Carbon Capture and Sequestration Technologies Program," at sequestration.mit.edu.

9. Some sources also include monitoring the CO_2 reservoir for leakage as a fourth criterion. "CO_2 Capture and Storage," *EPRI Journal*, spring 2007, p. 9.

10. International Energy Agency (IEA), *Technology Roadmap: Carbon Capture and Storage* (Paris: 2009).

11. Global CCS Institute, *The Status of CCS Overview 2010* (Canberra, Australia: October 2010).

12. Worldwatch calculations based on Global CCS Institute, op. cit. note 3.

13. Ibid.

14. Jon Gibbins and Hannah Chalmers, "Carbon Capture and Storage," *Energy Policy*, vol. 36 (2008), pp. 4,317–22.

15. Global CCS Institute, op. cit. note 3.

16. U.S. Department of Energy, "Enhanced Oil Recovery/CO_2 Injection," at www.fossil.energy.gov/programs/oilgas/eor, viewed 14 March 2011.

17. Global CCS Institute, op. cit. note 3.

18. IEA, op. cit. note 10.

19. Ibid.

20. Ibid.

21. "In Salah," at www.insalahco2.com, viewed 22 March 2011; Global CCS Institute, op. cit. note 3.

22. Global CCS Institute, op. cit. note 3.

23. Matthias Finkenrath, *Cost and Performance of Carbon Dioxide Capture from Power Generation, 2011*, Working Paper (Paris: IEA, 2010).

24. The Intergovernmental Panel on Climate Change writes, "With appropriate site selection…, a monitoring programme to detect problems, a regulatory system and the appropriate use of remediation methods to stop or control CO2 releases if they arise, the local health, safety and environment risks of geological storage would be comparable to the risks of current activities such as natural gas storage, EOR and deep underground disposal of acid gas." Intergovernmental Panel on Climate Change, *Carbon Dioxide Capture and Storage: Summary for Policymakers* (Geneva: September 2005), p. 12.

25. "Carbon Capture Project Leaking in to their Land,

Couple Says," (Toronto) *Globe and Mail*, 11 January 2011.

26. "Alleged Leaks from Carbon Storage Project Questioned," (Toronto) *Globe and Mail*, 13 January 2011.

27. On increased water usage and fuel consumption, see National Energy Technology Laboratory, *Carbon Capture Approaches for Natural Gas Combined Cycle Systems* (Washington, DC: U.S. Department of Energy, 2010); on water contamination, see Mark Little and Robert Jackson, "Potential Impacts of Leakage from Deep CO_2 Geosequestration on Overlying Freshwater Aquifers," *Environmental Science and Technology*, vol. 44 (2010), pp. 9,225–32.

28. Global CCS Institute, op. cit. note 1.

World Grain Production Down in 2010, but Recovering (pages 62–64)

1. "Staple Foods: What Do People Eat?" in Tony Loftas, ed., *Dimensions of Need, An Atlas of Food and Agriculture* (Rome: U.N. Food and Agriculture Organization (FAO), 1995).

2. FAO, *Food Outlook, Global Market Analysis*, June 2011; U.S. Department of Agriculture (USDA), "World Agricultural Production," August 2011 (wheat production).

3. FAO, op. cit. note 2; USDA, op. cit. note 2; FAO, "Global Cereal Production Forecast to Rise, But Food Insecurity to Continue," press release (New York: 6 October 2011).

4. FAO, *FAOSTAT Statistical Database*, at faostat.fao.org.

5. Ibid.

6. Ibid.

7. Ibid.

8. Rising irrigation from "Prospects by Major Sector Crop Production," in FAO, *World Agriculture: Toward 2015/2030. Summary Report* (Rome: 2002); fertilizers from Jonna P. Estudillo and Keijiro Otsuka, *Lessons from Three Decades of Green Revolution in the Philippines* (Japan: Foundation for Advanced Studies on International Development, November 2002); environmental degradation from Hira Jhamtani, "The Green Revolution in Asia: Lessons for Africa," in Lim Li Ching, Sue Edwards, and Nadia El-Hage Scialabba, eds., *Climate Change and Food Systems Resilience in Sub-Saharan Africa* (Rome: FAO, 2011), pp. 45–57.

9. FAO, op. cit. note 4.

10. Ibid.

11. World population data from U.N. Department of Economic and Social Affairs, Population Division, at esa.un.org/unpd/wpp.

12. World Health Organization, *Global and Regional Food Consumption Patterns and Trends*, at www.who.int/nutrition/topics/3_foodconsumption/en/index.html.

13. FAO, op. cit. note 2, p. 10.

14. Maxim Tkachenko, "Russia to Lift Grain Export Ban on July 1," *CNN World*, 28 May 2011.

15. U.S. Census Bureau, *The 2011 Statistical Abstract* (Washington, DC: 2011), Table 1374: Wheat, Rice, and Corn—Exports and Imports of Leading Countries: 2000 to 2009.

16. USDA, "Global Crop Production Review, 2010," at www.usda.gov; Whitney McFerron and Greg Quinn, "Canada's Wheat Crop to Shrink 17% from Last Year on Flooding in Prairies," *Bloomberg News*, 4 October 2010.

17. International Grains Council, *Report for Fiscal Year 2009/10* (London: January 2011).

18. USDA, op. cit. note 16.

19. U.S. Grains Council, *Corn*, at www.grains.org/corn.

20. Daniel O'Brien, *World Corn Supply—Demand Trends: An "At Risk" Position for MY 2011/12*, at www.agmanager.info.

21. Steven Wallander, Roger Claassen, and Cynthia Nickerson, *The Ethanol Decade: An Expansion of U.S. Corn Production, 2000–09* (Washington, DC: USDA, Economic Research Service, August 2011), p. 2.

22. Cited in ibid.

23. Ibid.

24. University of Southern California Sea Grant, *El Niño Facts*, at www.usc.edu/org/seagrant/Education/elnino/index.html.

25. USDA, op. cit. note 2, p. 23.

26. Ibid.

27. U.S. Census Bureau, op. cit. note 15.

28. FAO, *FAO Food Price Index*, at www.fao.org; Daniel Morgan, "Russia's Wheat Embargo Likely to Raise Global Food Costs," *The Fiscal Times*, 26 September 2010.

29. FAO, op. cit. note 28.

30. Loek Boonekamp, "Food Prices, the Grain of Truth," *OECD Observer*, May-June 2008.

31. FAO, op. cit. note 28.

32. FAO, *OECD–FAO Agricultural Outlook 2011–2020* (Rome: 2011).

33. W. E. Easterling et al., "Food, Fiber and Forest Products," in Intergovernmental Panel on Climate Change, *Climate Change 2007: Impacts, Adaptation*

and Vulnerability. Contribution of Working Group II to the Fourth Assessment Report of the Intergovernmental Panel on Climate Change (Cambridge, U.K.: Cambridge University Press, 2007), pp. 273–313.

34. David B. Lobell et al., "Climate Trends and Global Crop Production Since 1980," *Science*, 5 May 2011, pp. 616–20.

35. Ibid.

36. Ibid.

37. Sheeran quoted in Debora MacKenzie, "Calls for a Food-aid Revolution in a Post-surplus World," *New Scientist*, 5 June 2008.

Organic Agriculture Sustained through Economic Crisis (pages 65–68)

1. Helga Willer, "Organic Agriculture Worldwide – The Results of the FiBL/IFOAM Survey," in Helga Willer and Lukas Kilcher, eds., *The World of Organic Agriculture: Statistics and Emerging Trends 2011* (Bonn, Germany, and Frick, Switzerland: International Federation of Organic Agriculture Movements (IFOAM) and Forschungsinstitut für biologischen Landbau [Research Institute of Organic Agriculture], 2011), p. 41.

2. Willer and Kilcher, op. cit. note 1, p. 251.

3. International Service for the Acquisition of Agri-Biotech Applications (ISAAA), "Global Status of Commercialized Biotech/GM Crops 2010, Executive Summary," at www.isaaa.org; U.N. Food and Agriculture Organization (FAO), *FAOSTAT Statistical Database*, at faostat.fao.org, viewed 7 March 2011. Percentage based on the ratio of ISAAA's report of 1 billion hectares of agricultural land planted in genetically modified organism crops worldwide in 2010 to FAOSTAT's result of 48,836,977,000 hectares of agricultural land worldwide in 2008.

4. IFOAM, "Definition of Organic Agriculture," at www.ifoam.org.

5. Willer and Kilcher, op. cit. note 1, p. 246. France, Germany, Italy, and the United Kingdom were subtracted from the European Union total since these countries are also stand-alone G-20 members.

6. Willer and Kilcher, op. cit. note 1, p. 234.

7. Helga Willer, "Organic Agriculture in Europe: Overview," in Willer and Kilcher, op. cit. note 1, p. 153.

8. Ibid.

9. Willer and Kilcher, op. cit. note 1, p. 251.

10. Ibid.

11. Barbara Fitch Haumann, "North American Review,"

in Willer and Kilcher, op. cit. note 1, p. 201. U.S. and Canadian dollars were at parity in 2010.

12. Willer and Kilcher, op. cit. note 1, pp. 106, 122, 183, and 225.

13. Ibid., p. 247.

14. Ibid., p. 262; Hervé Bouagnimbeck, "Organic Farming in Africa," in Willer and Kilcher, op. cit. note 1, p. 106.

15. International Assessment of Agricultural Knowledge, Science and Technology for Development, *Agriculture at a Crossroads* (Washington, DC: 2009).

16. Office of the United Nations High Commissioner for Human Rights, "Right to Food: 'Agroecology Outperforms Large-scale Industrial Farming for Global Food Security,' says UN Expert," press release (Brussels: 22 June 2010).

17. Ibid.

18. Ibid.; Danielle Nierenberg, "In Ethiopia, Learning from Past Mistakes," *Nourishing the Planet* (blog), 2 November 2009.

19. Willer and Kilcher, op. cit. note 1, p. 245.

20. Salvador V. Garibay, Roberto Ugas, and Patricia Flores Escudero, "Organic Farming in Latin America and the Caribbean," in Willer and Kilcher, op. cit. note 1, p. 182.

21. Willer and Kilcher, op. cit. note 1, p. 249.

22. Garibay, Ugas, and Escudero, op. cit. note 20, p. 187. Third Country status, per Commission Regulation (EU) No. 471/2010, refers to "the arrangements for imports of organic products from third countries," having "established a list of third countries whose system of production and control measures for organic production of agricultural products are recognized as equivalent to those laid down" in Article 33(2) of Regulation (EC) No. 834/2007, Annex III to Commission Regulation (EC) No. 1235/2008. There are nine such countries with Third Country status: Argentina, Australia, Costa Rica, Japan, India, Israel, New Zealand, Switzerland, and Tunisia.

23. Garibay, Ugas, and Escudero, op. cit. note 20, p. 188.

24. Ibid., p. 183.

25. Ibid.

26. Ibid., p. 182.

27. Ibid., p. 183.

28. Willer and Kilcher, op. cit. note 1, p. 245.

29. Ong Kung Wai, "Organic Asia 2010," in Willer and Kilcher, op. cit. note 1, p. 122.

30. Willer and Kilcher, op. cit. note 1, p. 251.

31. Ibid., p. 256.

32. Ibid., p. 245.
33. "Global Sales of Organic Food & Drink Recovering," *Organic Monitor*, at www.organicmonitor.com/700 340.htm, viewed 3 March 2011.
34. Amarjit Sahota, "The Global Market for Organic Food & Drink," in Willer and Kilcher, op. cit. note 1, p. 62.
35. Ibid.
36. Ibid.
37. Diana Schaack, Helga Willer, and Susanne Padel, "The Organic Market in Europe," in Willer and Kilcher, op. cit. note 1, p. 157.
38. Ibid.
39. Organic Trade Association (OTA), *The Organic Trade Association's 2010 Organic Industry Survey (Summary)* (Brattleboro, VT: 2010).
40. "NBJ's Latest Market Research Shows U.S. Sales of Organic Foods and Beverages Slowed Considerably in 2008," *Nutrition Business Journal*, 20 March 2009.
41. Marsha Laux, "Organic Food Trends Profile," Agricultural Marketing Resource Center, Iowa State University, May 2010, at www.agmrc.org.
42. Economic Research Service, U.S. Department of Agriculture, "Organic Industry Statistics and Projected Growth," September 2010, at www.ers .usda.gov/data/organic, viewed 3 March 2011.
43. OTA and *KIWI Magazine*, *U.S. Families' Organic Attitudes & Beliefs Study, Executive Summary* (Greenfield, MA: OTA, June 2009).
44. Ibid. For a fuller discussion on the hidden costs in conventional foods, see Bryan Walsh, "Getting Real About the High Price of Cheap Food," *Time*, 31 August 2009, or the documentary *Food, Inc.*
45. Sahota, op. cit. note 34, p. 65.
46. Sophia Twarog, "Facilitating Global Organic Market Access." in Willer and Kilcher, op. cit. note 1, p. 76. FAO, IFOAM, and the U.N. Conference on Trade and Development are involved in the GOMA project.
47. Kung Wai, op. cit. note 29, p. 122.
48. Ibid., p. 123.
49. Karen Mapusua, "The Pacific Islands," in Willer and Kilcher, op. cit. note 1, p. 223.
50. Ibid., p. 224.
51. "Outsourcing's Third Wave," *The Economist*, 21 May 2009.
52. Savills plc, *Spotlight on International Farmland Markets* (London: January 2011), p. 2.
53. World Bank, *Agriculture and Rural Development & Development Economics Joint Notes, Land Policy*

Administration, Issue 54 (Washington, DC: 2010; Joachim von Braun and Ruth Meinzen-Dick, *"Land Grabbing" by Foreign Investors in Developing Countries*, Policy Brief No. 13 (Washington, DC: International Food Policy Research Institute, 2009).
54. OTA and *KIWI Magazine*, op. cit. note 43.

Sugar Production Dips (pages 69–71)

1. U.N. Food and Agriculture Organization (FAO), *FAOSTAT Statistical Database*, at faostat.fao.org.
2. Ibid.
3. Ibid.
4. Ibid.
5. Ibid.
6. Ibid.
7. Michael McConnell, Erik Dohlman, and Stephen Haley, "World Sugar Price Volatility Intensified by Market and Policy Factors," *Amber Waves* (U.S. Department of Agriculture), September 2010.
8. Ibid.
9. Ibid.
10. FAO, op. cit. note 1.
11. McConnell, Dohlman, and Haley, op. cit. note 7.
12. Ibid.
13. Ibid.
14. Ibid.
15. Ibid.
16. Ibid.
17. REN21, *Renewables 2010 Global Status Report* (Paris: 2010), p. 24.
18. Peter Zuurbier and Jos van de Vooren, *Sugarcane Ethanol: Contributions to Climate Change Mitigation and the Environment* (Wageningen: Wageningen Academic Publishers, 2008), pp. 56–57.
19. McConnell, Dohlman, and Haley, op. cit. note 7.
20. Zuurbier and van de Vooren, op. cit. note 18.
21. Bonsucro, "Bonsucro EU Production Standard," at www.bonsucro.com.

Fish Production from Aquaculture Rises While Marine Fish Stocks Continue to Decline (pages 72–75)

1. U.N. Food and Agriculture Organization (FAO), *The State of World Fisheries and Aquaculture 2010* (Rome: 2011); all 2009 data are provisional estimates.
2. Ibid.
3. FAO, *Food Outlook*, November 2010; 2010 statistics are November 2010 forecast.

4. FAO, *FAOSTAT Statistical Database*, at faostat.fao.org, updated March 2010.

5. Ibid.

6. Ibid.; FAO, op. cit. note 1.

7. FAO, op. cit. note 1.

8. Ibid.

9. Ibid.

10. FAO, *Fish Consumption Reaches All-time High*, press release (Rome: 31 January 2011).

11. FAO, op. cit. note 1.

12. Ibid.

13. FAO, *The State of World Fisheries and Aquaculture 2006* (Rome: 2007).

14. FAO, op. cit. note 1.

15. Ibid.

16. Ibid.

17. Ibid.

18. Ibid.

19. Ibid.

20. Ibid.

21. Ibid.

22. FAO, op. cit. note 13.

23. FAO, op. cit. note 1.

24. Ibid.

25. Dane Klinger and Kimiko Narita, "Peak Tuna," *Foreign Policy*, 12 February 2010.

26. Ibid.

27. World Wide Fund for Nature (WWF), "Meet the Amazing Atlantic Bluefin Tuna (*Thunnus thynnus*)," at wwf.panda.org; FAO, op. cit. note 1.

28. Pew Environment Group, "BP Oil Spill Threatens Bluefin Tuna Spawning Ground in Gulf of Mexico," at www.pewenvironment.org, 14 May 2010; U.S. Department of the Interior, "Strickland Announces Continued United States Support for International Proposal to Protect Bluefin Tuna," press release (Washington, DC: 3 March 2010).

29. WWF, "Bleak Future for Bluefin as Tuna Commission Only Marginally Trims Catches," press release (Paris: 27 November 2010).

30. Pew Environment Group, op. cit. note 28.

31. Ibid.

32. Margot L. Stiles et al., *Hungry Oceans: What Happens When the Prey is Gone?* (Washington, DC: Oceana, 2009).

33. Monterrey Bay Aquarium, "Aquaculture: Ensuring that All is Well Down on the Fish Farm," at www.montereybayaquarium.org/cr/cr_seafoodwatch/issues/aquaculture.aspx.

34. WWF, *Living Planet Report 2010: Biodiversity, Bioca-* *pacity, and Development* (Gland, Switzerland: 2010).

35. FAO, op. cit. note 1.

36. Ibid.

37. Ibid.

38. Ibid.

39. Ibid.

40. Ibid.

41. Ibid.

42. Ibid.

43. Ibid.

44. Ibid.

45. FAO, *Increasing the Contribution of Small-scale Fisheries to Poverty Alleviation and Food Security*, Technical Guidelines for Responsible Fisheries No. 10 (Rome: 2005); David K. Schor, *Artisanal Fishing: Promoting Poverty Reduction and Community Development through New WTO Rules on Fisheries Subsidies* (Geneva: U.N. Environment Programme, November 2005), p. 1; FAO, op. cit. note 1.

46. FAO, op. cit. note 45.

47. Ibid.

48. John W. Miller, "Global Fishing Trade Depletes African Waters: Poor Nations Get Cash; The Rich Send Trawlers; A Dearth of Octopus," *Wall Street Journal*, 23 July 2007.

49. Ibid.

50. WWF, "Depletion of Fisheries Could Affect Millions in West Africa," at wwf.panda.org.

51. Miller, op. cit. note 48.

52. Ibid.

53. FAO, op. cit. note 10.

54. "N. Zealand Urges US-Australia on Pacific Fisheries," *Agence France-Presse*, 21 February 2011.

55. Ibid.

56. FAO, op. cit. note 1; Crawford Allan, "CITES: Failure or Veiled Success?" press release (Gland, Switzerland: WWF, 13 April 2010).

57. WWF, op. cit. note 29.

58. Nicolas Gutiérrez, Ray Hilborn, and Omar Defeo, "Leadership, Social Capital and Incentives Promote Successful Fisheries," *Nature*, 17 February 2011, pp. 386–89.

59. Ibid.

60. Brian Halweil and Danielle Nierenberg, "Charting a New Path to Eliminating Hunger," in Worldwatch Institute, *State of the World 2011* (New York: W. W. Norton & Company, 2011), p. 3.

61. Ibid.

62. National Marine Protected Areas Center, "Frequently Asked Questions," at www.mpa.gov.

63. Ibid.

64. WWF, *Marine Protected Areas: Providing a Future for Fish and People* (Gland, Switzerland: undated).

65. Gutiérrez, Hilborn, and Defeo, op. cit. note 58.

66. FAO, op. cit. note 1.

67. Jae-Young Ko, *The Economic Value of Ecosystem Services Provided by the Galveston Bay/Estuary System* (Webster, TX: Texas Commission on Environmental Quality, August 2007).

Meat Production and Consumption Continue to Grow (pages 76–78)

1. U.N. Food and Agriculture Organization (FAO), "Meat and Meat Products," *Food Outlook*, June 2011.

2. Environmental Working Group (EWG), *Meat Eater's Guide to Climate Change and Health* (Washington, DC: July 2011).

3. Organisation for Economic Co-operation and Development (OECD) and FAO, "Meat," *OECD and FAO Agricultural Outlook 2011–2020* (Rome: June 2011).

4. FAO, op. cit. note 1.

5. Ibid.

6. Ibid.

7. Ibid.

8. Ibid.

9. Ibid.

10. Ibid.; WATT AgNet.com, "US Pig Meat Exports Up for 2011," 23 June 2011, at www.wattagnet.com.

11. FAO, op. cit. note 1.

12. Ibid.; U.S. Department of Agriculture (USDA), *Broiler Meat Summary Selected Countries*, April 2011, at www.fas.usda.gov.

13. FAO, op. cit. note 1.

14. Ibid.

15. Ibid.

16. Antonio Rota and Olaf Thieme, "The Livestock Challenge," *Rural 21*, June 2009, pp. 8–11.

17. UNESCO, Scientific Committee on Problems of the Environment, and U.N. Environment Programme, "Livestock in a Changing Landscape," *Policy Briefs No. 6*, April 2008.

18. D. P. Van Vuuren et al., "Outlook on Agricultural Change and its Drivers," *Agriculture at a Crossroads* (Washington, DC: 2009), pp. 255–305.

19. WWF, *Living Planet Report 2008* (London: 2008), p. 19.

20. Don Peden et al., "Water and Livestock for Human Development," in *Water for Wood, Water for Life: A Comprehensive Assessment of Water Management in Agriculture* (London and Colombo, Sri Lanka: Earthscan and International Water Management Institute, 2007), pp. 91–145.

21. FAO, *Livestock's Long Shadow, Environmental Issues and Options* (Rome: 2007), pp. xx and xxi.

22. EWG, op. cit. note 2.

23. Ibid.

24. FAO, op. cit. note 1.

25. B. Perry and K. Sones, *Global Livestock Disease Dynamics Over the Last Quarter Century: Drivers, Impacts and Implications* (Rome: FAO, 2009).

26. Eric Claas et al., "Human Influenza A H5N1 Virus Related to a Highly Pathogenic Avian Influenza Virus," *The Lancet*, February 1998, pp. 472–77; J. S. Malik Peiris, Leo L. M. Poon, and Yi Guan, "Emergence of a Novel Swine-Origin Influenza A Virus (S-OIV) H1N1Virus in Humans," *Journal of Clinical Virology*, July 2009, pp. 169–73; Henry Prempeh, Robert Smith, and Berit Müller, "Foot and Mouth Disease: The Human Consequences," *British Medical Journal*, March 2011, pp. 565–66; Paul Brown, "Bovine Spongiform Encephalopathy and Variant Creutzfeldt-Jakob disease," *British Medical Journal*, May 2001, pp. 841–44; Tariq A. Madani et al., "Rift Valley Fever Epidemic in Saudi Arabia: Epidemiological, Clinical, and Laboratory Characteristics," *Clinical Infectious Diseases*, October 2003, pp. 1,084–92.

27. James Newcomb, "Economic Impact of Selected Infectious Diseases," *Bio Economic Research Associates*, 2009.

28. Margaret Mellon, Charles Benbrook, and Karen Lutz Benbrook, *Hogging It! Estimates of Antimicrobials Abuse in Livestock* (Washington, DC: Union of Concerned Scientists, 2001).

29. Ralph Loglisci and David Love, "New FDA Numbers Reveal Food Animals Consume Lion's Share of Antibiotics," Center for a Livable Future, December 2010, at www.livablefutureblog.com.

30. J. C. Chee-Sanford et al., "Fate and Transport of Antibiotic Residues and Antibiotic Resistance Genes Following Land Applications of Manure Waste," *Journal of Environmental Quality*, April 2009, pp. 1,086–108.

31. Ibid.

32. Holly Dolliver, Kuldip Kumar, and Satish Gupta, "Sulfamethazine Uptake by Plants from Manure-Amended Soil," *Journal of Environmental Quality*, vol. 36, no. 4 (2007), pp. 1,224–30.

33. Rashmi Sinha et al., "Meat Intake and Mortality: A Prospective Study of Over Half a Million People," *Archives of Internal Medicine*, vol. 169, no. 6 (2009), pp. 562–71.

34. Kate Clancy, *Greener Pastures: How Grass-fed Beef and Milk Contribute to Healthy Eating* (Cambridge, MA: Union of Concerned Scientists, March 2006), pp. 1–4.

35. EWG, op. cit. note 2.

36. Ibid.

World's Forests Continue to Fall as Demand for Food and Land Goes Up (pages 80–82)

1. U.N. Food and Agriculture Organization (FAO), *Global Forest Resources Assessment 2010* (Rome: 2010). One square kilometer = 100 hectares = 247 acres.

2. Ibid.

3. Ibid.

4. Ibid.

5. Ibid.

6. Ibid.

7. Ibid.

8. FAO, *FAOSTAT: Land*, at faostat.fao.org, viewed 24 January 2011. Data compiled and grouped by Worldwatch Institute.

9. Ibid.

10. FAO, op. cit. note 1.

11. Ibid.

12. Ibid.

13. European Environment Agency, *The European Environment Assessment: State and Outlook 2010—Land Use* (Copenhagen: 2011).

14. FAO, op. cit. note 1.

15. N. Ramankutty and J. A. Foley, "Estimating Historical Changes in Global Land Cover: Croplands from 1700 to 1992," *Global Biogeochemical Cycles*, vol. 13 (1999), pp. 997–1027, data from Center for Sustainability and the Global Environment, "Global Land Use Data," at www.sage.wisc.edu, viewed 24 January 2011.

16. Ibid.

17. Organisation for Economic Co-operation and Development, *OECD Environmental Outlook to 2030* (Paris: 2008).

18. U.N. Population Division, *World Population Prospects: The 2008 Revision*, at esa.un.org/unpp, viewed 26 January 2011.

19. Earth Policy Institute, "Corn Production and Use for Fuel Ethanol in the United States, 1980–2010," at www.earth-policy.org, viewed 26 January 2011.

20. Conference of the Parties to the Convention on Biological Diversity, *Strategic Plan for Biodiversity 2011–2020*, COP 10 Decision X/2, at www.cbd.int, viewed 26 January 2011.

Tropical Forests Push Payments for Ecosystem Services onto the Global Stage (pages 83–85)

1. Millennium Ecosystem Assessment, *Ecosystems and Human Well-being: Synthesis* (Washington, DC: Island Press, 2005).

2. Tracy Stanton et al., *State of Watershed Payments: An Emerging Marketplace* (Washington, DC: Ecosystem Marketplace and Forest Trends, 2010); Becca Madsen et al., *Update: State of Biodiversity Markets* (Washington, DC: Forest Trends, 2011); Katherine Hamilton et al., *State of the Forest Carbon Markets 2009* (Washington, DC: Ecosystem Marketplace, 2009).

3. Stanton et al., op. cit. note 2.

4. Ibid.

5. Ibid.

6. Becca Madsen et al., *State of Biodiversity Markets Report: Offset and Compensation Programs Worldwide* (Washington, DC: Ecosystem Marketplace, 2010).

7. Madsen et al., op. cit. note 2.

8. Madsen et al., op. cit. note 6.

9. Ibid.

10. Hamilton et al., op. cit. note 2.

11. Ibid.

12. "Voluntary Carbon Market Climbs on Trees," *Carbon Positive*, 3 June 2011.

13. S. Wunder, *Payments for Ecosystem Services: Some Nuts and Bolts*, Occasional Paper No. 42 (Bogor, Indonesia: Center for International Forestry Research, 2005).

14. United Nations Framework Convention on Climate Change, "CDM FAQs," at cdm.unfccc.int.

15. James Salzman, *A Policy Maker's Guide to Designing Payments for Ecosystem Services*, Duke Law Faculty Scholarship, Paper 2081 (Durham. NC: Duke Law School, 2009).

16. Jie Li et al., "Rural Household Income and Inequality under the Sloping Land Conversion Program in Western China," *Proceedings of the National Academy of Sciences*, 25 April 2011.

17. Sara Scherr et al., *Developing Future Ecosystem Service Payments in China: Lessons Learned from Interna-*

tional Experience (Washington, DC: Forest Trends, 2006).

18. Wunder, op. cit. note 13.

19. U.S. Environmental Protection Agency, "Mitigation Banking Factsheet," at www.epa.gov.

20. For 12 percent, see G. R. van der Werf et al., "CO_2 Emissions from Forest Loss," *Nature Geoscience*, vol. 2 (2009), pp. 737–38; for 20 percent, see B. Metz et al., eds., *Contribution of Working Group III to the Fourth Assessment Report of the Intergovernmental Panel on Climate Change* (Cambridge, U.K.: Cambridge University Press, 2007).

21. Arild Angelsen and Sheila Wertz-Kanounnikof, "What Are the Key Design Issues for REDD and the Criteria for Assessing Options?" in Arild Angelsen, ed., *Moving Ahead with REDD Issues, Options and Implications* (Bogor, Indonesia: Center for International Forestry Research, 2008), pp. 11–21.

22. U.N. Food and Agriculture Organization, "Funding Gaps for Climate Change Adaptation a Threat to Food Supplies," press release (Rome: 3 December 2010).

23. "California Proposes Delaying Carbon Market a Year," *Reuters*, 29 June 2011.

24. Janet Lawrence, "Norway, Germany Give $90 Mln to Slow Deforestation," *Reuters*, 21 June 2011.

Value of Fossil Fuel Subsidies Declines, National Bans Emerging (pages 86–89)

1. International Energy Agency (IEA) *World Energy Outlook 2010* (Paris: 2010), "Executive Summary."

2. Ibid.

3. Ibid., p. 570.

4. Ibid., p. 571.

5. IEA et al., *Analysis of the Scope of Energy Subsidies and Suggestions for the G-20 Initiative* (Paris: June 2010); Table 1 and Figure 1 from "Fossil Fuel Consumption Subsidy Rates as a Proportion of the Full Cost of Supply, 2009," *World Energy Outlook 2010*, at www.iea.org/subsidy/index.html.

6. IEA et al., op. cit. note 5.

7. IEA, op. cit. note 1.

8. Global Subsidies Initiative, "Relative Subsidies to Energy Sources: GSI Estimates," Geneva, April 2010.

9. The IEA estimates that in 2009, government support for electricity from renewables and biofuels worldwide totaled $57 billion. See IEA, op. cit. note 1.

10. Bank Information Center, "World Bank Group Energy Sector Financing Update," Washington, DC, November 2010.

11. Ibid.

12. Ibid.

13. Global Subsidies Initiative, op. cit. note 8.

14. REN21, *2010 Global Status Report* (Paris: 2010).

15. IEA, op. cit. note 1.

16. "Arbeitsplätze durch Erneuerbare Energien," press release (Berlin: Federal Ministry for Environment, Nature Conservation and Nuclear Safety (BMU), 7 October 2010).

17. BMU, *Nationaler Energieeffizienzplan* (Berlin: October 2008).

18. Ibid.

19. Green Budget Germany, "Der Beitrag einer Modernen Umweltpolitik zur Haushaltskonsolidierung," Berlin, June 2010.

20. Friends of the Earth et al., *Green Scissors 2010* (Washington, DC: 2010).

21. World Bank, *Cost of Pollution in China: Economic Estimates of Physical Damages* (Washington, DC: February 2007), pp. xvii and 67. The 5.8 percent of GDP is for 2003.

22. Nicholas Stern, *The Economics of Climate Change: The Stern Review* (Cambridge, U.K.: Cambridge University Press, 2007).

23. "Total primary energy demand (TPED) represents domestic demand only and is broken down into power generation, other energy sector and total final consumption"; IEA, *World Energy Outlook 2010* (Paris: 2010), p. 707.

24. IEA, op. cit. note 23, p. 583f.

25. Global Subsidies Initiative, op. cit. note 8.

26. IEA et al., op. cit. note 5.

27. G20 Seoul Summit 2010, "The Seoul Summit Document," at www.interaction.org; "Leaders' Statement: The Pittsburgh Summit," 24–25 September 2009, at www.blogs.state.gov.

28. European Commission, *Energy 2020: A Strategy for Competitive, Sustainable and Secure Energy* (Brussels: November 2010).

29. Council of the European Union, "Council Decision on State Aid to Facilitate the Closure of Uncompetitive Coal Mines," Brussels, 9 December 2010.

30. U.S. Department of Energy, "Funding Highlights," at www.whitehouse.gov/sites/default/files/omb /budget/fy2011/assets/energy.pdf.

31. Global Subsidies Initiative, *Case Study: Lessons Learned from Indonesia's Attempts to Reform Fossil Fuel Subsidies* (Geneva: October 2010).

32. Ibid.

33. Vikas Bajaj, "India Cuts Subsidies for Fuel," *New York Times*, 25 June 2010.

34. Office of the Press Secretary, The White House, "The G-20 Summit in Toronto: Acting on Our Global Energy and Climate Change Challenges," press release (Washington, DC: 27 June 2010).

35. IEA, "Recent Developments in Energy Subsidies," *World Energy Outlook*, at www.worldenergyoutlook.org/Files/ann_plans_phaseout.pdf.

36. Ibid.

37. Ibid.

38. Ibid.

39. Ibid.

40. Ibid.

Energy Intensity Is Rising Slightly (pages 90–92)

1. World Bank, *GDP (constant 2000 US $)*, at worldbank.org; BP, *BP Statistical Review of World Energy* (London: June 2011).

2. World Bank, op. cit. note 1; BP, op. cit. note 1.

3. World Bank, op. cit. note 1; BP, op. cit. note 1.

4. World Bank, op. cit. note 1; BP, op. cit. note 1.

5. Tang Wee Liang, ed., *Entrepreneurship and Innovation in the Knowledge-based Economy: Challenges and Strategies* (Taipei: Asian Productivity Organization, July 2002).

6. World Bank, op. cit. note 1; BP, op. cit. note 1.

7. World Bank, op. cit. note 1; BP, op. cit. note 1.

8. World Bank, *GDP (constant 2005 international $)*, at worldbank.org; BP, op. cit. note 1.

9. World Bank, op. cit. note 8; BP, op. cit. note 1.

10. U.S. Department of Energy, Energy Information Administration, *World Crude Oil Prices*, at www.eia.gov.

11. Ibid.

12. World Bank, op. cit. note 8; BP, op. cit. note 1.

13. Joshua Zumbrun, "The Most Energy-Efficient Countries," *Forbes.com*, 7 July 2008.

14. World Bank, op. cit. note 8; BP, op. cit. note 1.

15. World Bank, op. cit. note 8; BP, op. cit. note 1.

16. World Bank, op. cit. note 8; BP, op. cit. note 1.

17. World Bank, op. cit. note 8; BP, op. cit. note 1.

18. "China's Stimulus Package: A Breakdown of Spending," *Economic Observer*, 7 March 2009.

19. World Bank, op. cit. note 8; BP, op. cit. note 1.

20. World Bank, op. cit. note 8; BP, op. cit. note 1.

World Labor Force Growing at Divergent Rates (pages 94–97)

1. U.N. Population Division, Department of Economic and Social Affairs, *World Population Prospects: The 2010 Revision* (New York: 2011).

2. Ibid.

3. International Labor Organization (ILO), *Global Employment Trends 2011* (Geneva: 2011). In addition, over 20 percent of workers subsist at an extreme poverty level of $1.25 per day.

4. U.N. Population Division, op. cit. note 1.

5. Ibid.

6. Ibid.; World Bank, "Gross Domestic Product 2010," at worldbank.org.

7. U.N. Population Division, op. cit. note 1; World Bank, op. cit. note 6.

8. ILO, "EAPEP: Economically Active Population Estimates and Projections 1980–2020," at laborsta.ilo.org.

9. Ibid.

10. World Bank, *2011 World Development Indicators* (Washington, DC: 2011).

11. D. Bloom and J. Williamson, "Demographic Transitions and Economic Miracles in Emerging Asia," *World Bank Economic Review*, vol. 12 (1998), pp. 419–56; D. Bloom, D. Canning, and P. Malaney, "Demographic Change and Economic Growth in Asia," *Population and Development Review*, vol. 26 Suppl. (2000), pp. 257–90.

12. H. Boulhol, *The Effects of Population Structure on Employment and Productivity*, Economics Department Working Papers No. 684 (Paris: Organisation for Economic Co-operation and Development (OECD), 2009); W. Lutz, J. Crespo Cuaresma, and W. Sanderson, "The Demography of Educational Attainment and Economic Growth," *Science*, 22 February 2008, pp. 1,047–48.

13. U.N. Population Division, op. cit. note 1.

14. The range of projections cited here is based on the "medium" and "low" fertility variants in U.N. Population Division, op. cit. note 1. The assumptions underlying the two variants lead to a difference in projected fertility rates of one-half child per woman by the 2020s. The U.N. also produces a "high" fertility variant that assumes that the average global fertility rate would immediately begin rising from current levels; most demographers would consider this unlikely.

15. U.N. Population Division, op. cit. note 1.

16. Ibid.
17. Ibid.
18. Ibid.
19. Ibid.
20. Ibid.
21. Ibid.
22. Ibid.
23. Uganda Population Secretariat, Ministry of Finance, Planning and Economic Development, *Uganda: Population Factors & National Development* (Kampala: 2010).
24. Ibid.
25. U.N. Population Division, op. cit. note 1.
26. Ibid. The United Nations estimates Japan's fertility rate was 1.32 in 2005–10. Official government data show a slightly higher rate of 1.37 in 2008–09 (Japan Statistics Bureau, *Statistical Handbook of Japan 2010*, at www.stat.go.jp/english).
27. U.N. Population Division, op. cit. note 1.
28. Ibid.
29. Ibid.
30. Ibid.
31. Ibid.
32. Ibid. The U.N. estimates India's fertility rate was 2.73 in 2005–10 while official government data show a lower fertility rate of 2.6 in 2008–09 (Office of Registrar General, "Maternal & Child Mortality and Total Fertility Rates: Sample Registration System," at www.censusindia.gov.in).
33. U.N. Population Division, op. cit. note 1.
34. Ibid.
35. Ibid.
36. World Bank, op. cit. note 10.
37. Ibid.
38. U.N. Population Division, *Trends in International Migrant Stock: The 2008 Revision* (New York: 2009).
39. Ibid.
40. Ibid.
41. Ibid.
42. U.N. Population Division, op. cit. note 1.
43. ILO, op. cit. note 8.
44. Ibid.
45. Available from OECD, "Ageing and Employment Policies—Statistics on Average Effective Age of Retirement," at www.oecd.org.
46. Ibid.
47. V. Skirbekk, "Age and Productivity Capacity: Descriptions, Causes and Policy Options," *Ageing Horizons*, Issue no. 8 (2008), pp. 4–12.

Women Slowly Close Gender Gap with Men (pages 98–100)

1. Unless indicated otherwise, rankings and data in this article are from Ricardo Hausmann, Laura D. Tyson, and Saadia Zahidi, *The Global Gender Gap Report 2010* (Geneva: World Economic Forum, 2010).
2. "Sex Ratio," in Central Intelligence Agency, *The World Factbook*, at www.cia.gov/library/publications/the-world-factbook/fields/2018.html.
3. Margaret C. Hogan et al., "Maternal Mortality for 181 Countries, 1980–2008: A Systematic Analysis of Progress towards Millennium Development Goal 5," *The Lancet*, 8 May 2010, pp. 1,609–23.
4. Klaus Schwab, ed., *The Global Competitiveness Report 2010–2011* (Geneva: World Economic Forum, 2010).
5. World Databank, *World Development Indicators & Global Development Finance*, at data.worldbank.org.
6. U.N. Population Division, *World Population Prospects: The 2008 Revision, Population Database*, at esa.un.org/unpp, viewed 19 January 2011.

Numbers of Overweight on the Rise (pages 101–03)

1. Worldwatch population numbers, based on data for 2005 and 2010, are from the U.N. Population Division, *World Population Prospects: The 2008 Revision* (New York: 2008); 2002 data interpolated by the U.N. Population Division using 2000 and 2005 figures.
2. U.N. Population Division, op. cit. note 1.
3. See, for example, National Heart Lung and Blood Institute (NHLBI), "What Are the Health Risks of Overweight and Obesity?" at www.nhlbi.nih.gov.
4. Body mass index (BMI) data from World Health Organization (WHO), *Global Infobase*, at www.apps.who.int/infobase/Comparisons.aspx.
5. Japan defines obesity as a BMI of 25 or higher.
6. U.N. Population Division, op. cit. note 1.
7. Ibid.
8. Ibid.
9. Ibid.
10. Ibid.
11. "Poor" and "wealthy" were defined using the International Monetary Fund's Gross Domestic Product at Purchasing Power Parity per Capita index for 2010, at www.imf.org; data for Cuba, Micronesia, North Korea, and Somalia from Central Intelligence

Agency, *The World Factbook*, at www.cia.gov/library/publications/the-world-factbook.

12. U.N. Population Division, op. cit. note 1; WHO, op. cit. note 4.

13. The 2010 estimate for Canada was 63 percent, a good 15 percent lower than in the United States. In 2002 Canada's rate was 60 percent. Its increase of 3 percentage points in eight years is less than half that of the United States, which jumped 7 percentage points.

14. The r-value is 0.4387.

15. Children in Japan are becoming heavier, however; see Yumi Matsushita et al., "Trends in Childhood Obesity in Japan over the Last 25 Years from the National Nutrition Survey," *Obesity Research*, vol. 12 (2004), pp. 205–14.

16. U.N. Population Division, op. cit. note 1; WHO, op. cit. note 4.

17. Ibid.

18. Ibid. Data for the United States are substantially higher than reported by the Centers for Disease Control and Prevention, likely because the figures here consider adults to be those over 15.

19. Nearby Fiji and Vanuatu were also unusually high, at 58 and 64 percent, respectively.

20. U.N. Population Division, op. cit. note 1; WHO, op. cit. note 4.

21. See, for example, Ricardo Uauy, Cecelia Albala, and Juliana Kain, "Obesity Trends in Latin America: Transiting from Under- to Overweight," *The Journal of Nutrition*, vol. 131 (2001), pp. 893S–99S.

22. For a summary, see, for example, Centers for Disease Control and Prevention, "U.S. Obesity Trends,"

at www.cdc.gov. The prevalence of highest rates in the South is striking.

23. See Stephen T. McGarvey, "Obesity in Samoa and a Perspective on its Etiology in Polynesians," *American Journal of Clinical Nutrition*, June 1991, pp. 1586S–94S. The author noted that Samoans who move to Hawaii also tend to become much heavier. Genetic factors are suggestive, but the literature is not conclusive; see Karolina Åberg et al., "Susceptibility Loci for Adiposity Phenotypes on 8p, 9p, and 16q in American Samoa and Samoa," *Obesity*, vol. 17, no. 3 (2009), pp. 518–24.

24. U.N. Population Division, op. cit. note 1; WHO, op. cit. note 4.

25. U.N. Population Division, op. cit. note 1; WHO, op. cit. note 4.

26. U.N. Population Division, op. cit. note 1; WHO, op. cit. note 4.

27. Although drugs may control AIDS-related Wasting Syndrome and weight loss from the disease in general, HIV-positive persons remain at serious risk; see AIDSMEDS, "Wasting Syndrome," at www.aidsmeds.com/articles/Wasting_6934.shtml.

28. The author thanks Lindsay Nauen for this observation.

29. H. Salome Kruger et al., "Obesity in South Africa: Challenges for Government and Health Professionals," *Public Health Nutrition*, vol. 8, no. 5 (2005), pp. 491–500.

30. Anne Case and Alicia Menendez, "Sex Differences in Obesity Rates in Poor Countries: Evidence from South Africa," *Economics and Human Biology*, vol. 7, no. 3 (2009), pp. 271–82.

31. NHLBI, op. cit. note 3.

The Vital Signs Series

Some topics are included each year in *Vital Signs*; others are covered only in certain years. The following is a list of topics covered in *Vital Signs* thus far, with the year or years they appeared indicated in parentheses. The reference to 2006 indicates *Vital Signs 2006–2007*; 2007 refers to *Vital Signs 2007–2008*.

ENERGY AND TRANSPORTATION

Fossil Fuels
  Carbon Use (1993)
  Coal (1993–96, 1998, 2009, 2011)
  Fossil Fuels Combined (1997, 1999–2003, 2005–07, 2010)
  Natural Gas (1992, 1994–96, 1998, 2011–12)
  Oil (1992–96, 1998, 2009, 2012)
Renewables, Efficiency, Other Sources
  Biofuels (2005–07, 2009–12)
  Biomass Energy (1999)
  Combined Heat and Power (2009)
  Compact Fluorescent Lamps (1993–96, 1998–2000, 2002, 2009)
  Efficiency (1992, 2002, 2006)
  Geothermal Power (1993, 1997)
  Hydroelectric Power (1993, 1998, 2006, 2012)
  Nuclear Power (1992–2003, 2005–07, 2009, 2011–12)
  Solar Power (1992–2002, 2005–07, 2009–12)
  Solar Thermal Power (2010)
  Wind Power (1992–2003, 2005–07, 2009–12)

Transportation
  Air Travel (1993, 1999, 2005–07, 2011)
  Bicycles (1992–2003, 2005–07, 2009)
  Car-sharing (2002, 2006)
  Electric Cars (1997)
  Gas Prices (2001)
  High-Speed Raid (2012)
  Motorbikes (1998)
  Railroads (2002)
  Urban Transportation (1999, 2001)
  Vehicles (1992–2003, 2005–07, 2009–12)

ENVIRONMENT AND CLIMATE

Atmosphere and Climate
  Carbon and Temperature Combined (2003, 2005–07, 2009–10)
  Carbon Capture and Storage (2012)
  Carbon Emissions (1992, 1994–2002, 2009)
  CFC Production (1992–96, 1998, 2002)
  Global Temperature (1992–2002)
  Ozone Layer (1997, 2007)
  Sea Level Rise (2003, 2011)
  Weather-related Disasters (1996–2001, 2003, 2005–07, 2009–11)

Natural Resources, Animals, Plants
  Amphibians (1995, 2000)
  Aquatic Species (1996, 2002)
  Birds (1992, 1994, 2001, 2003, 2006)
  Coral Reefs (1994, 2001, 2006, 2010)
  Dams (1995)
  Ecosystem Conversion (1997)
  Energy Productivity (1994, 2012)
  Forests (1992, 1994–98, 2002, 2005–06, 2012)
  Groundwater (2000, 2006)
  Ice Melting (2000, 2005)
  Invasive Species (2007)
  Mammals (2005)
  Mangroves (2006)
  Marine Mammals (1993)
  Organic Waste Reuse (1998)
  Plant Diversity (2006)
  Primates (1997)
  Terrestrial Biodiversity (2007, 2011)
  Threatened Species (2007)
  Tree Plantations (1998)
  Vertebrates (1998)
  Water Scarcity (1993, 2001–02, 2010)
  Water Tables (1995, 2000)
  Wetlands (2001, 2005)
Pollution
  Acid Rain (1998)
  Air Pollution (1993, 1999, 2005)
  Algal Blooms (1999)
  Hazardous Wastes (2002)
  Lead in Gasoline (1995)
  Mercury (2006)
  Nuclear Waste (1992, 1995)
  Ocean (2007)
  Oil Spills (2002)
  Pollution Control Markets (1998)
  Sulfur and Nitrogen Emissions (1994–97)
Other Environmental Topics
  Bottled Water (2007, 2011)
  Energy Poverty (2012)
  Environmental Indicators (2006)
  Environmental Treaties (1995, 1996, 2000, 2002)
  Protected Areas (2010)
  Semiconductor Impacts (2002)
  Transboundary Parks (2002)
  World Heritage Sites (2003)

FOOD AND AGRICULTURE

Agriculture
  Farmland Quality (2002)
  Fertilizer Use (1992–2001, 2011)
  Genetically Modified Crops (1999–2002, 2009)
  Grain Area (1992–93, 1996–97, 1999–2000)
  Irrigation (1992, 1994, 1996–99, 2002, 2007, 2010)
  Nitrogen Fixation (1998)
  Organic Agriculture (1996, 2000, 2010, 2012)
  Pesticide Control or Trade (1996, 2000, 2002, 2006)
  Pesticide Resistance (1994, 1999)
  Soil Erosion (1992, 1995)
  Urban Agriculture (1997)
Food Trends
  Aquaculture (1994, 1996, 1998, 2002, 2005)
  Aquaculture and Fish Harvest Combined (2006–07, 2009–12)
  Cocoa Production (2002, 2011)
  Coffee (2001)
  Eggs (2007)
  Fish Harvest (1992–2000)
  Grain Production (1992–2003, 2005–07, 2009–12)
  Grain Stocks (1992–99)
  Grain Used for Feed (1993, 1995–96)
  Livestock (2001)
  Meat (1992–2000, 2003, 2005–07, 2009–12)
  Milk (2001)
  Soybeans (1992–2001, 2007)
  Sugar and Sweetener Use (2002, 2012)

GLOBAL ECONOMY AND RESOURCES

Resource Economics
  Agricultural Subsidies (2003)
  Aluminum (2001, 2006–07)
  Arms and Grain Trade (1992)
  Commodity Prices (2001)
  Fossil Fuel Subsidies (1998, 2012)
  Gold (1994, 2000, 2007)
  Illegal Drugs (2003)
  Materials Use (2011)
  Metals Exploration (1998, 2002)
  Metals Production (2002, 2010)
  Paper (1993–94, 1998–2000)
  Paper Recycling (1994, 1998, 2000)
  Payments for Ecosystem Services (2012)
  Roundwood (1994, 1997, 1999, 2002,
    2006–07, 2011)
  Steel (1993, 1996, 2005–07)
  Steel Recycling (1992, 1995)
  Subsidies for Environmental Harm (1997)
  Wheat/Oil Exchange Rate (1992–93, 2001)
World Economy and Finance
  Agribusiness (2007)
  Agricultural Trade (2001)
  Aid for Sustainable Development (1997, 2002)
  Carbon Markets (2009, 2012)
  Developing-Country Debt (1992–95,
    1999–2003)
  Environmental Taxes (1996, 1998, 2000)
  Food Aid (1997)
  Global Economy (1992–2003, 2005–07,
    2009–11)
  Green Jobs (2000, 2009)
  Microcredit (2001, 2009)
  Private Finance in Third World (1996,
    1998, 2005)
  R&D Expenditures (1997)
  Seafood Prices (1993)
  Socially Responsible Investing (2001, 2005,
    2007)
  Stock Markets (2001)
  Trade (1993–96, 1998–2000, 2002, 2005)

  Transnational Corporations (1999–2000)
  U.N. Finances (1998–99, 2001)
Other Economic Topics
  Advertising (1993, 1999, 2003, 2006, 2010)
  Charitable Donations (2002)
  Child Labor (2007)
  Cigarette Taxes (1993, 1995, 1998)
  Corporate Responsibility (2006)
  Cruise Industry (2002)
  Ecolabeling (2002)
  Government Corruption (1999, 2003)
  Informal Economies (2007)
  Labor Force (2010, 2012)
  Nanotechnology (2006)
  Pay Levels (2003)
  Pharmaceutical Industry (2001)
  PVC Plastic (2001)
  Satellite Monitoring (2000)
  Television (1995)
  Tourism (2000, 2003, 2005)
  Unemployment (1999, 2005, 2011)

POPULATION AND SOCIETY

Communications
  Computer Production and Use (1995)
  Internet (1998–2000, 2002)
  Internet and Telephones Combined (2003,
    2006–07, 2011)
  Satellites (1998–99)
  Telephones (1998–2000, 2002)
Health
  AIDS/HIV Incidence (1994–2003, 2005–07)
  Alternative Medicine (2003)
  Asthma (2002)
  Avian Flu (2007)
  Breast and Prostate Cancer (1995)
  Child Mortality (1993, 2009)
  Cigarettes (1992–2001, 2003, 2005)
  Drug Resistance (2001)
  Endocrine Disrupters (2000)
  Fast-Food Use (1999)
  Food Safety (2002)

Health Aid Funding (2010)
Health Care Spending (2001)
Hunger (1995, 2011)
Immunizations (1994)
Infant Mortality (1992, 2006)
Infectious Diseases (1996)
Life Expectancy (1994, 1999)
Malaria (2001, 2007)
Malnutrition (1999)
Mental Health (2002)
Mortality Causes (2003)
Noncommunicable Diseases (1997)
Obesity (2001, 2006, 2012)
Polio (1999)
Sanitation (1995, 1998, 2006, 2010)
Soda Consumption (2002)
Traffic Accidents (1994)
Tuberculosis (2000)

Military
Armed Forces (1997)
Arms Production (1997)
Arms Trade (1994)
Landmines (1996, 2002)
Military Expenditures (1992, 1998, 2003, 2005–06)
Nuclear Arsenal (1992–96, 1999, 2001, 2005, 2007)
Peacekeeping Expenditures (1994–2003, 2005–07, 2009)
Resource Wars (2003)
Wars (1995, 1998–2003, 2005–07)
Small Arms (1998–99)

Reproductive Health and Women's Status
Family Planning Access (1992)
Female Education (1998)
Fertility Rates (1993)
Gender Gap (2012)
Maternal Mortality (1992, 1997, 2003)
Population Growth (1992–2003, 2005–07, 2009–11)
Sperm Count (1999, 2007)
Violence Against Women (1996, 2002)
Women in Politics (1995, 2000)

Other Social Topics
Aging Populations (1997)
Educational Levels (2011)
Homelessness (1995)
Income Distribution or Poverty (1992, 1995, 1997, 2002–03, 2010)
Language Extinction (1997, 2001, 2006)
Literacy (1993, 2001, 2007)
International Criminal Court (2003)
Millennium Development Goals (2005, 2007)
Nongovernmental Organizations (1999)
Orphans Due to AIDS Deaths (2003)
Prison Populations (2000)
Public Policy Networks (2005)
Quality of Life (2006)
Refugees (1993–2000, 2001, 2003, 2005)
Refugees-Environmental (2009)
Religious Environmentalism (2001)
Slums (2006)
Social Security (2001)
Sustainable Communities (2007)
Teacher Supply (2002)
Urbanization (1995–96, 1998, 2000, 2002, 2007)
Voter Turnouts (1996, 2002)

Worldwatch Reports

On Climate Change, Energy, and Materials

184: Powering the Low-Carbon Economy, 2010
183: Population, Climate Change, and Women's Lives, 2010
182: Renewable Energy and Energy Efficiency in China, 2010
180: Red, White, Green: A New U.S. Approach to Biofuels, 2009
179: Mitigating Climate Change Through Food and Land Use, 2009
178: Low-Carbon Energy: A Roadmap, 2008
175: Powering China's Development: The Role of Renewable Energy, 2007
169: Mainstreaming Renewable Energy in the 21st Century, 2004
160: Reading the Weathervane: Climate Policy From Rio to Johannesburg, 2002
157: Hydrogen Futures: Toward a Sustainable Energy System, 2001
151: Micropower: The Next Electrical Era, 2000
149: Paper Cuts: Recovering the Paper Landscape, 1999

On Ecological and Human Health

181: Global Environmental Change: The Threat to Human Health, 2009
174: Oceans in Peril: Protecting Marine Biodiversity, 2007
165: Winged Messengers: The Decline of Birds, 2003
153: Why Poison Ourselves: A Precautionary Approach to Synthetic Chemicals, 2000

On Economics, Institutions, and Security

186: Creating Sustainable Prosperity in the United States, 2011
185: Green Economy and Green Jobs in China: Current Status and Potentials for 2020, 2011
177: Green Jobs: Working for People and the Environment, 2008
173: Beyond Disasters: Creating Opportunities for Peace, 2007
168: Venture Capitalism for a Tropical Forest: Cocoa in the Mata Atlântica, 2003
167: Sustainable Development for the Second World: Ukraine and the Nations in Transition, 2003
166: Purchasing Power: Harnessing Institutional Procurement for People and the Planet, 2003
164: Invoking the Spirit: Religion and Spirituality in the Quest for a Sustainable World, 2002
162: The Anatomy of Resource Wars, 2002
159: Traveling Light: New Paths for International Tourism, 2001
158: Unnatural Disasters, 2001

On Food, Water, Population, and Urbanization

176: Farming Fish for the Future, 2008
172: Catch of the Day: Choosing Seafood for Healthier Oceans, 2006
171: Happier Meals: Rethinking the Global Meat Industry, 2005
170: Liquid Assets: The Critical Need to Safeguard Freshwater Ecosytems, 2005
163: Home Grown: The Case for Local Food in a Global Market, 2002
161: Correcting Gender Myopia: Gender Equity, Women's Welfare, and the Environment, 2002
156: City Limits: Putting the Brakes on Sprawl, 2001
154: Deep Trouble: The Hidden Threat of Groundwater Pollution, 2000
150: Underfed and Overfed: The Global Epidemic of Malnutrition, 2000

To see our complete list of Reports, visit www.worldwatch.org/taxonomy/term/40

Price of each Report is $12.95 plus shipping and handling.